Poisoned Science

(The 1960s Corruption of Scientific Methods For Careers and Causes)

The Paradigm Company, Boise, Idaho

The Paradigm Company, Boise, Idaho
(208) 322-4440

email: paradigm@srnrl.com
web: www. paradigmphysics.com
        www.srnrl.com

First Printing, July 2016

ISBN: 978-1533645081

Other books by Lawrence Dawson
•The Quantum Dimension
•Four Dimensional Atomic Structure
•The Death of Reality (2015 edition)
•The Quantum Dimensional review of Newton and Einstein
•Edge-of-Universe Photo Confirms Quantum-Dimensional Cosmology over
  Big Bang

Poisoned Science

*(The 1960s Corruption of Scientific Methods
For Careers and Causes)*

Lawrence Dawson

DEDICATED TO THE HUMAN RACE
whose intellectual capacities to reason have been so debased by the deceptive science documented in this book.

INTRODUCTION: *"What the hell does 'Poisoned Science' mean to me if I don't give a damn about science?"*

I could tell you that you have been sold a *"fairytale image"* of the cosmos and everything within it by highly celebrated scientists, but you are likely to reply, *"So what? What has that to do with me and my everyday life?"*

It might be better to begin by telling you why it matters to you. The *"fairytale cosmos"*— that *"fairytale"* in which you believe because men *"smarter"* than you told you it was true; belief in that *"fairytale"* has left many without reason and without the ability to see "facts," even when those facts slap them in the face. Many of you and your contemporaries have devolved to become *"monkey people"* with only a fraction of the intellectual prowess of your forefathers.

You were told by the celebrated scientists, and you believed it, that, in the far distant past, everything you see in the sky— the stars, the sun, the moon— were compressed into a space smaller than a pin prick. Over many ages this pin-prick, which contained everything in the universe, expanded to a circumference of 2 billion light years— which was changed in the 1960s to 14 billion light years. These distances are so immense that we don't have words in our language to describe them in miles. The pinprick with an unmeasurably small circumference and with everything in it, expanded to 82,300 billion, billion miles (82 thousand three hundred *times* a billion *times* a billion miles).

Your great-great grandfather might have had logical questions about this claim, questions about the "Big Bang" which would have scandalized you today. You have been told that *"all the smart people"* believe in the *"Big Bang."* Your great-great grandfather would not have shared this belief because, unlike you, he did not consider "smart" to be an honor conferred by a *"troupe of chattering monkeys"* but something which an

i

individual earned through the application of reason and logic. Your great-great grandfather might have asked *"If everything is crammed into such a small size, then exploded to become so huge does that mean things are becoming thinner and thinner?"* The question would have been suggested by his familiarity with explosions which he knew reduced densities or *"thinned things out."*

One can almost hear your thoughts, *"Please, great-great grandpa don't be so dumb. Scientists can understand these things but you can't!"* You would be wrong. There is not a higher logic for "scientists" and lower version for "great-great grandfathers." The question which great-great grandfather may have put to the "Big Bang" also disturbs scientists because it is a valid logic.

Scientist have tried to answer great-great grandfather's question with *"flimflam."* Contemporary scientists would tell great-great grandfather that all matter is being held together and not allowed to become less dense by an alleged increased gravitational influence from *"dark matter."* This influence purportedly holds matter together while the surrounding space expands around it. This alleged *"dark matter"* is "invisible" and has supposedly 4 *times* greater mass than does "visible" matter.[1] Your ancestor would have rejected this argument as inconsistent with experience since neither he, nor anyone he knew, had ever encountered objects which cannot be seen but which have "weight."

One of the great differences between your generation and that of your great-great grandfather is the way that *"non-observable facts"* are established. For you and your generation such *"facts"* are established by socially shared beliefs, the acceptance of which supplies social rewards. For great-great grandfather, however, a *"non-observable fact"* must be objectively referenced. He wouldn't have credit as a *"fact"* a meteorite falling in China simply because everyone believes it to be true. He would have needed to read an eye-witness report in order to credit it as a *"non-observed fact."*

A cultural revolution has occurred. It has replaced objective referrals

---

1   *"Dark matter makes up 80% of the Universe—but where is it all?"* Ars Technica
      http://arstechnica.com/features/2014/07/dark-matter-makes-up-80-of-the-universe-but-where-is-it-all/

for our *"non-observed facts"* with accreditation by *"social belief."* By the 1960s this transformation had introduced a debasement to the hard sciences.

For example, in the 1960s the mathematical formula underpinning the "Big Bang" theory was changed— without reference to any data at all— in order to make the time since the beginning of the alleged expansion conform with other measures of the age of the universe. The new formula and its longer age of the universe did not refer to data and was accepted by consensus and by consensus alone.

During the same time frame, a new field of particle physics was established, *"exo-data"* (outside data), in order to include unproven particle theories in the same classification system with the known particles from nuclear physics. In fact the new particle physics and its artificial classification system required suppression of previously acquired knowledge about nuclear particles (protons and neutrons). Both of these scientific debasements, as well as the rational corrections[2], are covered in detail in this book. These poisoned streams in cosmology and particle physics could have been eliminated if the correct theory had been recognized.

The *"Big Bang"* and *"particle physics"* debasements were accepted without valid scientific proofs and were established solely as a consensus belief system. This led to the corruption of other fields of science; specifically to the emergence of a politically popular environmentalism which was able to suppress contradicting scientific data in support of the popular beliefs.

However, the greatest damage caused by this transformation of science from objectively referenced categories and rigorous proofs to a socially enforced belief system may not have been to science itself . The greatest damage may have been to the general population, resulting in a loss of intellectual capacities. The public has been taught to short-circuit reason, to ignore facts, and to call the results *"intellegence."* In doing so,

---

2   The development of quantum-dimensional mathematics has corrected primitive quantum mechanics. Those mathematics also establish a PROVEN alternative to *"Big Bang"* cosmology and a more accurate model of nuclear and electron atomic structure. The quantum-dimensional solutions are fully referenced.

men have lost the capacity to defend themselves. They have been rendered intellectually incompetent and cannot take rational precautions in areas of fundamental human needs.

There are examples of the inabilities of our contemporary *"mental deficients"* to apply rudimentary logic in their self interest. The primary example is the assault upon fossil fuels by *"greenhouse gas"* proponents. Both power plants and automobiles are under assault as contributors to atmospheric carbon dioxide levels which are said to be increasing earth temperatures through an alleged *"greenhouse effect."*

The actual scientific logic governing this claim is rather simple and straightforward, but the general public has lost the capacity to evaluate using simple logic. Everything within and upon the earth gives off a form of thermal radiation which is related to its temperature. These thermal radiation outputs are infrared wavelengths which are identified as "planetary thermal wavelengths" by NASA[3]. During the night, every-thing's temperature drops as its daytime heat is radiated off. This radiation can bounce back to earth by atmospheric conditions such as clouds. This atmospheric retention or "bouncing back of thermal radiation" is called a *"greenhouse effect."*

The "greenhouse effect" is easily pinpointed in your own personal experience. If you leave your car in the summer sun with the windows rolled up, heat builds up in the interior because the sunlight is let in by the widows but the return thermal radiation from the interior is blocked. The fact that glass blocks thermal radiation frequencies is the principle behind an agricultural greenhouse.

The existence of *"radiation frosts"* demonstrate the differences between atmospheric vs. glass containment of thermal radiation. According to the Encyclopedia Britannica *"radiation frost occurs on clear nights ....when the outgoing radiation is excessive and the air temperature is not necessarily at the freezing point."*[4] It is a frost condition of concern to the fruit industry that occurs when lack of cloud cover causes plants and

---

3   National Aeronautics and Space Administration, Science Mission Directorate. (2010). *Infrared Waves.* From Mission:Science website: http://missionscience.nasa.gov/ems/07_infraredwaves.html
4   *https://www.britannica.com/science/radiation-frost*

the earth to thermally radiate into space, dropping plant temperatures to below freezing. The atmosphere itself does not block or *"reflect back"* thermal radiation wavelengths. Cloud cover (water condensation), however, does reflect thermal radiation. The atmosphere alone does not act like glass to produce a "greenhouse effect." The existence of condensed water droplets in the atmosphere do act like glass.

These natural observations of atmospheric interactions with thermal radiation do not give support to the environmental arguments about an atmospheric *"greenhouse effect."* There have been no reports of decreases in clear-night radiation frosts due to the build up of carbon dioxide in the atmosphere. If carbon dioxide is a *"greenhouse gas,"* as claimed for it, the major constituents of the atmosphere, oxygen and nitrogen, clearly are not. The "radiation frost" proves that to be the case. The percentage of the alleged *"greenhouse gas,"* CO_2, is too miniscule to have much of an effect.

Even after the alleged build up of atmospheric CO_2 from use of fossil fuels, atmospheric CO_2 is only one part *per* two thousand five hundred parts of air (0.04%). This trace amount of carbon dioxide— less than four hundredths of a percentage point— is necessary to sustain life on this planet. The atmospheric oxygen-carbon cycle provides the great energy symbiosis between plants and animals which sustains life. If the traces of carbon dioxide should disappear from the atmosphere, all plant life would die to be followed by the animals due to dextrose starvation; the dextrose sugar molecule produced by plant photosynthesis being the energy source required by all animal cells.

Your great-great grandfather's logic would never have allowed the claim that an atmospheric trace of four hundredths of a percentage point could produce a *"greenhouse effect"* heat buildup. His everyday experience would rule against it. If he had been able to carefully roll up car windows so that only "0.04%" of the window space were blocked by the glass, the interior of the car would not retain heat like a greenhouse.

However, you are "smarter" than your great-great grandfather because you have the words of *"great scientists"* which have infused you with *"scientific knowledge."* But how, exactly, has science contradicted great-

great grandfather's logic in this matter of atmospheric CO_2? How can a small atmospheric trace element, such as CO_2, initiate such alleged catastrophic climate change by producing an alleged *"greenhouse effect?"*

Remember our old friend *" the transformation of science from objectively referenced categories and rigorous proofs to a socially enforced belief system?"* The categorization of carbon dioxide as a *"greenhouse gas"* was removed from any *"objective referent"* and supplied a new socially constructed meaning. No tests were made to actually measure the amount of thermal radiance blockage which the atmospheric traces caused and how these actual measures changed when the traces of carbon dioxide increased. No *"objectively referred"* rigorous testing of atmospheric CO_2 thermal radiance blockage were conducted.

Instead of factually testing the *"greenhouse gas"* properties of the atmospheric traces of carbon dioxide, those properties were artificially assumed using computer modeling. The traces were supplied *"super blockage"* properties in computer models in order to explain a completely unrelated phenomenon. It had been discovered that the earth retains a portion of the energy supplied by solar radiation. That is, the earth doesn't radiate off all the heat energy supplied by the sun's radiation. The *"super blockage"* characteristic of the atmospheric traces of carbon dioxide were assumed to explain the earth's retention of solar supplied heat. Under this assumption, small changes in these traces would have massive consequences.[5]

However, the hypothesis that the earth's retention of solar supplied heat is explained by an assumed *"super thermal-radiation blockage"* from atmospheric traces of carbon dioxide has never been proved.

Further, it is founded upon an abysmal ignorance of the nature of thermal radiation. Those thermal radiation wavelengths which are characteristically released by the earth, are identified as *" planetary frequencies"* by NASA. It has been demonstrated experimentally that

---

5   *"If carbon dioxide makes up only a minute portion of the atmosphere, how can global warming be traced to it? And how can such a tiny amount of change produce such large effects?"* by Pieter Tans, a senior scientist at the National Oceanic and Atmospheric Administration (NOAA) Earth System Research Laboratory. Scientific American: http://www.scientificamerican.com/article/if-carbon-dioxide-makes-u/

"planetary thermal radiation frequencies" (given off by temperatures below 118° C or 244.4° Fahrenheit) retain a portion of their heat by the nuclear processes within the radiating atoms.[6] The removal of atmospheric carbon dioxide traces from an *"objectively referenced greenhouse gas"* category in order to explain earth retention of solar supplied heat was completely unnecessary. The retention is explained completely by thermal radiation characteristics themselves. A "false science" has been substituted for scientific ignorance.

In science, you may have been too quick to deserted the logical capacities of your great-great grandfather and to adopt artificially constructed belief systems which claim to be able to confer upon you the status of *"intelligence and knowledge."* The resultant loss of intellectual capacities may have spilled over into other, more personal areas of your life. The complete decay of moral reason within many areas of the collective life seem to indicate this is the case.

This book is dedicated to the reacquisition of the human capacity for logic and reason. Those areas of science which address *"non-observable facts"* are exactly those areas in which that capacity has been lost to irrational collective belief systems. The road back to logic and reason will not be an easy one. It will require a painful use of an atrophied intelligence used upon scientific complexities. I urge you to persevere for you are fighting for your own ability to appreciate reality in a world made complex by artificially imposed irrational scientific belief systems. To this end I have tried to write this book to an intelligent, high-school educated audience, not to a scientific one.

---

6   *"Secondary statistical analysis of MIT's thermoelectrically pumped 'over unity' LED study proves the nuclear magnetic current thermal signature"* in *Four Dimensional Atomic Structure,* p.p. 184-193. Dawson, L. The Paradigm Company, Second Edition Printing 2015. ISBN: 978-0941995351

1. Modernist Science as Mythology

Strong evidence indicates that there has been a tendency toward a fraudulent science since the 1960s. However, those whose attitudes have been formed by the popular culture will immediately react to this statement as "nonsense." The fraudulent science has been sold as the *"vox populi"* and not established by the scientific method. It has come to function as a belief required for membership in an artificially constructed *new cultural order.*

Those who have profited from the belief system— primarily accredited "scientists" in universities and government financed research institutions — will rage against this proposition. They will say that advances in science over the last 50 years disproves the "fraudulent science" charge. They will claim that advances such as those in medicine, especially medical imaging, and, most especially, the discoveries in semiconductors which have led to the computer revolution, prove that underlying physical theories cannot be "fraudulent."

Those advances, however, are all what Isaac Newton called *"practical mechanics"* in that they are only technological extensions of pre-1960s discoveries. They are not similar to Faraday's demonstration of electrical current induction by a magnetic field which initiated the invention of electrical motors and generators; nor of the Maxwell/Heaviside electromagnetic wave theory which initiated radio; nor of the Rutherford Group's atomic model which ultimately resulted in fission nuclear energy. None of the scientific advances of the last 50 years have sprung from post-1960s pure science.

Probably the most important technological advance in the last 50 years has been the miniaturization of computers which led to tremendous increases in power and which, in turn, made the small, personal computer possible. The miniaturization of computer circuitry was achieved by advances in semiconductors.

1

The fundamentals of the semiconductor advances required to miniaturize computer circuitry were all in place by the 1960s. The concept of *"band gap"* (Wilson 1931) *"bipolarity of current"* (Baedeker 1908), *"holes"* (Heisenberg and Peierls 1931), *"p-n junction"* (Davydov 1938), *"thermionic emission"* (Bethe 1942), *"bipolar transistor"* (Shockley 1945), *"integrated circuit"* (Kilby 1958), *"tunnel diode"* (Esaki 1957-58), *"light emitting diode"* (Holonyak 1961) were in place and guided the technological developments of miniaturized computer chips and LED screens.[1] Again these latter day scientific advances were practices of Newton's *"practical mechanics,"* not produced by advances in theoretical science or by pure research.

> *"At the inaugural International Solid-State Circuits Conference held on the campus of the University of Pennsylvania in Philadelphia in 1960, a young computer engineer named Douglas Engelbart introduced the electronics industry to the remarkably simple but groundbreaking concept of "scaling."*

> *"Dr. Engelbart, who would later help develop the computer mouse and other personal computing technologies, theorized that as electronic circuits were made smaller, their components would get faster, require less power and become cheaper to produce — all at an accelerating pace.*

> *"Sitting in the audience that day was Gordon Moore, who went on to help found the Intel Corporation, the world's largest chip maker. In 1965, Dr. Moore quantified the scaling principle and laid out what would have the impact of a computer-age Magna Carta. He predicted that the number of transistors that could be etched on a chip would double annually for at least a decade, leading to astronomical increases in computer power.*

> *"His prediction appeared in Electronics magazine in April 1965 and was later called Moore's Law. It was never a law of physics, but rather an observation about the economics of a young industry that ended up holding true for a half-century."[2]*

---

1   *"History of Semiconductors,"* Lukasiak, L and Jakubowski, A. Journal of Telecommunications and Information Technology, 1/2010
2   *"Smaller, Cheaper, Faster, Over: The Future of Computer Chips."* John Markoff, *The New York Times.* Sept. 26, 2015

However, to fully understanding the *"scaling"* of integrated circuits to gain computer miniaturization requires high levels of technical competence in pre-1960s semiconductor theory and practice. Semiconductor theory is incredibly complex and esoteric. It is beyond the general public, even the educated general public.

Further, the mathematics which establish the computer's possibilities are also beyond the general public. An "integrated circuit" consists of a number of transistors "stamped" upon a semiconductor "wafer" (primarily silicon). "Scaling" consists of increasing the number of transistors in the circuit by reducing their size. Each transistor is a switch with an "n" or "p" position. These bipolar transistor switches support the binomial numbering system which consists of two digits; "1" ("n" pole which establishes the electron-flow side) and "0" ("p" pole establishing the "hole" side).

Any number can be written with these two digits which is duplicated in the transistor's "p" and "n" junctions. Multiplication can be calculated by adding any number to itself for the number of times indicated by the multiplier. Division can be calculated as the number of times the denominator can be subtracted from the whole. The number of transistor "switches" on the wafer determine how rapidly these calculations can be made. Increasing the "scale" increases the speed of calculations and the number of calculations which can be made simultaneously (computer power).

To the calculation power available to the integrated circuit's transistor "p-n junction" switches was added a logical "if, then" function. "If" the calculation achieved a certain result, "then" a switch was opened which initiated another and an additional set of calculations. As the "scale" of the number of transistors etched upon the semi-conductor wafer increased, a seconds set of mathematics was applied. "Algorithmetrics" (or algorithms) is a mathematical science for the subdivision of a field which was introduced by the ancient Greek philosopher Euclide. As the number of transistor switches imprinted upon the semiconductor wafer increased, a means had to be found to isolate a portion of them to specific tasks. "Alogrithmetric" mathematics made this possible.[3] The number of

---

3 *"ALGORITHMS IN MODERN MATHEMATICS AND COMPUTER SCIENCE."* Knuth, Donald E., *Stanford University Dept. of Computer Science; Report STAN-CS-80-786, January 1980.*

transistors stamped upon a computer chip increased from "2.3 thousand" in 1971 to "2.6 billon" in 2011.[4]

In short, both the semiconductor technology and the required mathematical support for a revolution in computer miniaturization were in place prior to 1960. The computer revolution required no advances in either scientific theory or in pure scientific research to accomplish. It was a *"practical mechanic"* built upon a half-century of semiconductor discoveries and established mathematical principles.

Simultaneously with this "scaling" of bipolar transistors to miniaturize computer processors, there was a destructive pattern developing in theoretical physics. In the late 1950s and early 1960s three poisonous streams converged to create a fraudulent science. Scientific theory was disconnected from experimentation and observation and increasingly became validated solely by "peer agreement" or social consensus.

The first poison stream was provided by the Cambridge philosopher, Ludwig Wittgenstein, who deconstructed language with his 1953 book *"Philosophical Investigations."* Wittgenstein argued that words could not identify objective reality because words were only artificial and subjective impositions. He deconstructed language by removing our words as having *"objective referents"* and replaced *"word meaning"* as being only a subjectively imposed mental construct without any relationship to objective external reality. By the 1960s, Wittgenstein's linguistic deconstruction was undermining the scientific method in academic science departments[5].

The second science-poisoning stream was the late 1950s revision of Hubble's Constant in support of "Big Bang" cosmology. The theory proposes that the universe has been and is expanding from an original infinitesimally small "singularity point." The theory was built upon Edwin Hubble's 1929 data table which showed that stellar distances, as measured by Cepheids, directly correlated with the amount of redshift (shift

---

4   *"Microprocessor Transistor Counts 1971-2011"* Wikipedia;
     https://en.wikipedia.org/wiki/Transistor_count
5   *"Why I don't have a Ph.D. and, by extension— 'what does your Ph.D. actually
     mean?'"* Dawson, L. The Snake River N-Radiation Lab.
     http://www.paradigmphysics.com/phdata.pdf

downward in wavelength) in the stellar light. This "redshifting" he postulated was due to the Doppler Effect which indicated that stars at greater distance where receding from us at greater velocities. This is exactly the condition one would expect from an expanding universe following a "Big Bang." Hubble established a constant for recession which "best fit" his data table.

"Hubble's Constant" was recession velocity *per* unit of distance for any light source from us. Since "velocity *divided by* distance," mathematically, resolves to "1 *over* time," Hubble's constant actually identified the amount of time the universe had been expanding since the "Big Bang."[6]

Hubble's original constant resolves to approximately 2 billion years of expansion time . The constant *times* stellar distance from us gives recession velocity (velocity equals distance *over* time since expansion). The proposed recession velocity for any distance can be calculated using Hubble's Constant and the amount of redshift can be predicted by plugging recession velocity into the standard Doppler Effect formula.

A problem occurred almost immediately with Hubble's theory which proposed that the relationship between redshift and stellar distance was explained as "Big Bang" recession velocity. The constant which best fit his data table identified an an age of the universe as 2 billion years old. However, the half-lives of radioactive decay material in the earth's rocks gave a much older age. The ratio of the "lead" (Pb) decay-product to "uranium," for example, identified an earth age of 4 billion years. The "Big Bang" theory as applied to Hubble's data required an age of the universe be proposed. Yet this age was 2 billion years too short as compared to the age of the earth as measured by radioactive decay.

In science, if a theory doesn't fit all relevant data, then the theory should be abandoned to search for an alternative explanation. This was not done in the case of Hubble's "Big Bang." The theory was sustained in the face of contradicting geological data and despite the fact that it began to present other problems.

---

6   *"The Quantum Curvature of Space vs. an Expanding Universe,"* in *Edge of Universe Photo Confirms Quantum-Dimensional Cosmology over Big Bang,* p. 12. Dawson, L. The Paradigm Company, Boise, Idaho. Nov. 2015. ISBN: 9781519237132

The Hubble theory required that all geometric space be expanding, including matter, as well as the vacuum surrounding it. Matter would expand, becoming less dense, and orbital distances would increase causing orbital bodies to escape gravity. All sorts of theoretical "fixes" were proposed for the "Big Bang," such that matter might retain its density and orbital positions while the vacuum around it expanded.

The most important of these "fixes" was the concept of "dark matter" which increased gravitational influence to constrain visible matter from expanding. "Dark matter" was supposedly invisible.[7]

Further, Hubble's "Big Bang" and expanding universe was incompatible with Einstein's general theory of relativity and its required "cosmological constant."[8] Ten years before Hubble's "Big Bang," Albert Einstein's Relativity Field Equations had accurately predicted the amount of spacial curvature caused by the sun; the spacial curvature which had been measured by A. S. Eddington and F. W. Dyson during the solar eclipse of 1919.[9] Einstein's cosmological constant was a key mathematical component of the field equation which correctly predicted the Eddington/Dyson measurements.

The cosmological constant is a tension which Einstein postulated must exist in vacuum, a postulated tension which was used to calculate the spacial curvature measured by Eddington/Dyson ten years before Hubble's "Big Bang" theory. The "cosmological constant" is a vacuum tension which is shown to be the exact mathematical equivalent of the quantum-dimensional "time force" which sustains vacuous space.[10] However, this tension can only exist in a static universe. It is completely incompatible with Hubble's expanding universe.

"In 1917, Einstein was applying his new theory of general relativity to the structure of space and time. General relativity says that mass affects the shape of space and the flow of time. Gravity results because

---

7 *"Dark Energy, Dark Matter."* NASA Science: Astrophysics.
   http://science.nasa.gov/astrophysics/focus-areas/what-is-dark-energy/
8 *"The Expanding Universe and Hubble's Law"* Mastin, Luke
   http://www.physicsoftheuniverse.com/topics_bigbang_expanding.html
9 *The Quantum Dimensional Review of Einstein vs. Newton,* p.p. 5-12. Dawson, L. The
   Paradigm Company., Aug. 2015, ISBN: 978-1516918096
10 Ibid. p.p. 12-14

space is warped by mass. The greater the mass, the greater the warp.

"But Einstein, like all scientists at that time, did not know that the universe was expanding. He found that his equations didn't quite work for a static universe, so he threw in a hypothetical repulsive force that would fix the problem by balancing things out, an extra part that he called the 'cosmological constant.'

"Then, in the 1920s, astronomer Edwin Hubble, using a type of star called a Cepheid variable as a "standard candle" to measure distances to other galaxies, discovered that the universe was expanding. The idea of the expanding universe revolutionized astronomy. If the universe was expanding, it must at one time have been smaller. That concept led to the Big Bang theory, that the universe began as a tiny point that suddenly and swiftly expanded to create everything we know today.

"Once Einstein knew the universe was expanding, he discarded the cosmological constant as an unnecessary fudge factor. He later called it the "biggest blunder of his life," according to his fellow physicist George Gamow."[11]

Despite the fact that Einstein's *"vacuum tension"* cosmological constant had correctly predicted the 1919 measurement of the spacial curvature caused by the sun, Einstein abandoned and denounced the concept under the influence of Hubble's expanding universe. Once again, relevant data was ignored in support of the "Big Bang" theory.

The retention of Hubble's "Big Bang" explanation of his 1929 redshift/stellar-distance data in the face of contradictory data from the radioactive decay of earth rocks, as well as the 1919 spacial curvature measurements, proved unfortunate. Had the scientific method been properly applied and the "Big Bang" explanation of the redshift/distance data been deserted to search for an alternative, a much needed major scientific correction might have been discovered. The quantum-dimensional correction of primitive quantum mechanics might have been discovered in 1929.

The alternative explanation of Hubble's discovery of a direct relationship between redshift and stellar distance is a quantum curvature of

---

11 The Hubble Site:
   http://hubblesite.org/hubble_discoveries/dark_energy/de-did_einstein_predict.php

space as enforced by Einstein's proposed "vacuum tension." The greater the distance to the stellar light source, the greater the enforced curvature of space. Light would follow the curvature instead of the direct distance and wavelength would be "stretched" in direct proportion to the ratio of enforced curvature to direct distance. Science might have discovered the "quantum law of curvature" as the true explanation of Hubble's data.

The "quantum law of curvature" states that, as cosmological distances increase, those distances become more Euclidean and less quantum. Euclidean distances are "kinked" into curvature[12] and quantum distances are not. As distances decrease, these distances become proportionally more quantum and this proportion is reflected by the focal points of an ellipse. The focal points of an ellipse are made by splitting the center of a circle and spreading the two points an equal distance along a major axis. The space of separation between the two points is vacuum or a "quantum space" which plays no part in defining the curvature of the ellipse's periphery. The ratio of curvature distance to the straight distance along the major axis increases as the space of separation (quantum distance) between focal points decreases. Distance between focal points decreases to "0" and maximum elliptical curvature at the *"macro quantum distance."* Light from objects at greater distances than the *"macro quantum"* can no longer reach us as curvature drops it out of our sight.

An alternative explanation of Hubble's redshift-to-distance data could have been provided by the cosmological constant which was contained within Einstein's relativity field equations. Einstein's "vacuum tension" had been confirmed by the spacial curvature measurements made by Eddington/Dyson during the 1919 solar eclipse. If this vacuum tension were to increase as interstellar distances increased (as proposed by quantum-dimensional geometry), then this increased spacial curvature over interstellar distances could explain Hubble's data. The quantum curvature model, as applied to Hubble's data, actual explains more of the statistical variance between prediction and measurement than does Hubble's "Big Bang."[13]

---

12 *"Introduction to Quantum-Dimensional Geometry and the Relationship between Quantum-Squared Vacuum and a Strict Euclidean Definition of Vacuum,"* in *The Quantum Dimensional Review of Einstein vs. Newton,* p.p. 2-4. Op. Cit.

13 *Edge of Universe Photo Confirms Quantum-Dimensional Cosmology over Big Bang,* p. 28. Dawson, L. Op. Cit.

Despite its incompatibility with the radioactive decay rates in rocks and the Eddington/Dyson measurement of spacial curvature, Hubble's "Big Bang" expanding universe was embraced as an "article of faith" by the scientific community. As we have seen, the social consensus on the "Big Bang" was so strong as to pressure Einstein to denounce his "cosmological constant" despite its 1919 confirmation.

Even as the "Big Bang" became scientific dogma, scientists were increasingly uncomfortable with the 2 billion year time frame which Hubble needed to explain his data-set as Doppler Effect. While attempts had been made to "revise" Hubble's Cepheid distance measurements in order to get a longer age of the universe, no empirical data was applied to these revisions.[14] Hubble retained allegiance to his constant and its 2 billion year age until his death in 1953.

With Hubble's death, a frenzy to revise Hubble's Constant to make it compatible with a presumed older universe broke out. This frenzied revision was lead by Hubble's graduate student assistant, Allan Sandage. By 1958, Sandage had issued a series of speculative revisions to Hubble (not supported by published research data). In 1958, Sandage issued the modernist revision of Hubble's constant which persists to the present. This revision made the constant compatible with the presumed age of the universe (14 billion years) and was achieved by reducing Hubble's original constant by a factor of 86%.

The modernist "exo-data" (outside data) revision of Hubble's Constant can be tested against Hubble's original data set. The "exo-data revision" can calculate a presumed distance for the redshifts as measured by Hubble. Those distances are well beyond our detection capabilities for the "light pulses" by which Cepheids are measured. Hubble had originally estimated a "2.0 Mpc" (6.524 million light years) as the limit for which light pulses could be detected. All of the revised-constant distance predictions for the Hubble data set are beyond this limit with a "new" upper limit of "50.789 million light years."[15] This is well beyond the distance at which the small variations in light intensity which identify Cepheids might be detected.

---

14 *Edge of Universe Photo Confirms Quantum-Dimensional Cosmology over Big Bang,*
    p. 23-24. Op. Cit.
15 Ibid. p. 27

This is not to say that the Sandage "exo-data revision" was made without *argumentative support.* In fact, *"peer-reviewed argumentation" and "socially conferred honor"* replaced demonstration of the new constant value via data. Sandage would be rewarded with 6 "prestige" awards from 1957 to 1975, including a "National Medal of Science," which is something of a record in astronomy for "conferred prestige" and which is somewhat interesting because he is only credited with one small empirical discovery, the astroid "(96155) 1973 HA."[16] Sadage's 86% reduction in Hubble's Constant was rationalized by an "exo-data" assault upon the Shapley/Eddington Cepheid scale which Hubble had used to determine stellar distances in his 1929 data table.

In consultation via letter with Arthur Eddington (the same Eddington who would confirm Einstein's cosmological constant in 1919), the American astronomer Harlow Shapley had identified the physical characteristics of Cepheids. Cepheids are pulsating stars with their time of pulse identifying their absolute brightness. By measuring Cepheid pulses for 230 cases within globular star clusters, Shapley statistically developed a brightness-to-pulse scale which allowed him to measure distances to the Cepheids by comparing their absolute brightness to the brightness we receive from them. By these distance measurements, Shapley was able to make the first correct estimation of our own Milky Way galaxy's size.[17]

In 1924, Hubble contacted Shapley to help measure the distances to two Cepheids discovered in the nearby Andromeda galaxy. Shapley responded by *"promis[ing] to send a revised period-luminosity curve."[18]* This consultation demonstrated that Hubble was applying the Shapley/Eddington scale to the long distance Cepheid measurements (in Mega-parsecs) which he was making for the 1929 "Big Bang" data table.

The "exo-data" revision of the Shapley/Eddington Cepheid scale, in order to gain a longer "Big Bang" age, began in 1952 with Walter Baade. Baade proposed that the composition of stars controlled their luminosity and that "type I" stars were more metallic in composition and were brighter. He proposed that the Shapley-Eddington scale was calibrated on

16 *"Allan Sandage"* Wikipedia.   https://en.wikipedia.org/wiki/Allan_Sandage
17 *"Shapley, Harlow."* Complete Dictionary of Scientific Biography | COPYRIGHT 2008 Charles Scribner's Sons.
18 Ibid.

dimmer "type II" while the Hubble data was composed of the brighter "type I" and thus under-estimated distances hence the "Big Bang" age was understated.

Not only was the Baade proposed revision "exo-data" but it is statistically impossible. The Shapley/Eddington scale was statistically built upon 230 observations. If the population pool of Cepheids contains two types, as Baade proposed, it is statistically impossible that the Shapley sample of 230 would contain only "type II." Despite the fact that the Shapley/Eddington Cepheid scale was built statistically on a large sample and thus provided an immediate statistical test of the Baade claim, no such test was forthcoming. Rather, the Baade claim was accepted by fiat and the Hubble Constant time frame was doubled.

"In 1952 Baade surprised his fellow astronomers by announcing (at the 1952 Conference of the International Astronomical Union, in Rome) his determination of two separate populations of Cepheid variable stars in the Andromeda Galaxy, resulted in a doubling of the estimated age of the universe (from 1.8 to 3.6 billion years). Hubble had posited the earlier value; he had considered only the weaker Population II Cepheid variables as standard candles. After Baade's pronouncements, Sandage showed that astronomers' previous assumption, that the brightest stars in galaxies were of approximately equal inherent intensity, was mistaken in the case of H II regions which he found not to be stars and inherently brighter than the brightest stars in distant galaxies. This resulted in another 1.5 factor increase in the calculated age of the universe, to approximately 5.5 billion years.[19]*"* [20]

Baade's revision upward of the time of expansion since the "Big Bang" is statistically preposterous. His claim that Hubble had *"considered only the weaker Population II Cepheid variables as standard candles"* is an assault upon the Shapley-Eddington scale since that scale was Hubble's *"standard candles."* However, the Shapley survey of Cepheids was one of the most extensive ever conducted. Shapley's Cepheid survey had upped

---

19 *FOOTNOTE IN ORIGINAL.* "Singh, Simon (November 1, 2005). Big Bang: The Origin of the Universe. ISBN 978-0007162215."
20 *Allan Sandage"* Wikipedia. Op. Cit.

the original Leavitt survey by a factor of "five." Henreitta Leavitt had observed only 47 Cepheids by which she had first established that the brightness of a Cepheid is related to its periodicity.[21]

Shapley refined the original Leavitt Cepheid brightness-to-periodicity scale by taking a large enough Cepheid sample (230) to allow a statistical comparison of Cepheids of similar periodicity. His larger sample allowed him to statistically compare variations in measured light brightness (indicating distance) across categories to gain a more refined estimate of original brightness for any periodicity category. If the Baade theory is correct and stellar material divides stars into two populations— the "brighter" type I and the "dimmer" type II—then the Shapley categorization of his data by periodicity would cease to work. A mixture of different brightness within Shapley's periodicity categories would eliminate all statistical regularity. Baade knew this and proposed that the Shapley/Eddington scale from the 230 Cepheid samples (Hubble's *"standard candles"*) were all "type II."

The actual history of Baade's 1952 revision of Hubble's "Big Bang" age of the universe, from approximately 2 billion years to approximately 4 billion years, is disturbing. It cannot be explained as mathematical error or statistical malfeasance. It is deliberate deception designed to make the "Big Bang" cosmology coherent with the presumed age of the universe. The deceptive practice was furthered by Allan Sandage who took the lead in all further revisions of Hubble's "Big Bang" age of the universe, after Hubble died in 1953.

Sandage's first revision would be his last which made any appeal to data. Sandage's first post-Hubble revision increased the Baade revision by another 1.5 *times* to a "Big Bang" age of the universe of approximately 6 billion years. This revision was justified by alleging that the brightest stars were *"found not to be stars and [were] inherently brighter than the brightest stars in distant galaxies."*

However, the Shapley-Eddington scale was not built by comparing Cepheids to the *"brightest stars"* but by comparing Cepheids of equal periodicity. The brightest stars were not used in the scaling at all.

---

21 *Edge of Universe Photo Confirms Quantum-Dimensional Cosmology over Big Bang,*
   p. 18. Op. Cit.

Sandage made several further revisions, finally settling upon the 86% reduction of Hubble's original constant and its approximately 14 billion year old universe (the currently accepted value). All subsequent revisions were what the National Institute of Standards and Technology call *"time variations of the constant"* which are made without any *"observation of [the] time dependence of [the] constant which might be relevant to the recommended value."*[22] That is, Sandage's revisions were made without observational support yet are accepted by the National Institute of Standards and Technology as the legitimate value. They are accepted by consensus and consensus alone.

To summarize, in 1929 Edwin Hubble composed a table of 24 Cepheids for which he had measured distances as well as the measured redshift in their light.[23] He had used the Shapley-Eddington Cepheid scale to determine the astronomical distances to these Cepheids. The Shapley/Eddington scale was built upon the most extensive observations of Cepheids which had been conducted up to that point.

Hubble proposed that the redshift had resulted from the Doppler Effect produced by alleged recession velocities of a theoretical expanding universe. His data table indicated a direct relationship between redshift (recession velocities) and distance. From his data, he inductively constructed a constant which would explain the theoretical velocity-to-distance relationship. The constant consisted of a "velocity *per* astronomical distance" parameter which, when multiplied by a star's distance, would derive the recession velocity between ourselves and the star. His empirically derived constant was approximately "500 kilometers *per* second *per* mega-parsec[24]."

"Hubble's Constant," however, is actually a time measure as "velocity *over* distance *equals* the *inverse* of time." Hubble's Constant actually indicated the amount of time which the universe had been expanding, hence it identified the alleged age of the universe.

---

22 *Edge of Universe Photo Confirms Quantum-Dimensional Cosmology over Big Bang,* p.p. 15-16. Op. Cit.
23 Ibid. p. 19
24 Ibid.

The expanding-universe age which fit Hubble's data was incompatible with other scientific measures of cosmological age. The radioactive decay of uranium in earth rocks gave an age for the earth which was 2 billion years older than Hubble's age of the universe as given by his expanding universe explanation of the redshift-to-distance data for his Cepheid table.

The "Big Bang" expanding universe explanation of the data was retained as scientific dogma despite the fact that an alternative explanation of the redshift-to-distance data had been suggested by an astronomical measurement 10 years earlier. In 1919, A.S. Eddington, the same Eddington who had consulted with Harlow Shapley on his Cepheid scale, had measured the sun's curvature of space by photographing the displacement of stars around the sun's shoulder during a solar eclipse.[25] The photographs confirmed the prediction by Einstein's Field Equations that the curvature would be twice that predicted by gravitational curvature alone. The Einstein Field Equations incorporated a theoretical "vacuum tension" which Einstein called "the cosmological constant." The Eddington photos essentially confirmed the existence of Einstein's "cosmological constant."

Although not recognized at the time, Einstein's "vacuum tension," or cosmological constant, could provide an alternative explanation of the Hubble redshift-to-distance data. Recently it has been shown that Einstein's "cosmological constant" is the equivalent of the "quantum force" which composes an unrecognized extra quantum dimension to our spacial geometry.[26] Redshift of light could be caused by the quantum curvature of space enforced by Einstein's "vacuum tension." It is shown mathematically that the quantum curvature explanation of the redshift-to-distance relationship better fits Hubble's original data set.[27]

However, in 1929, no alternative explanation for the Cepheid data was sought and Hubble's "Big Bang" expanding universe was accepted by fiat, despite its age incompatibility with other scientific cosmological age measurements. Instead of recognizing his cosmological constant as a

---

25 *The Quantum Dimensional Review of Einstein vs. Newton,* p.p. 5-12. Dawson, L. Op. Cit.
26 Ibid. p.p.12-16
27 *Edge of Universe Photo Confirms Quantum-Dimensional Cosmology over Big Bang,* p. 28. Op. Cit.

possible alternative explanation of Hubble redshift-to-distance data, Einstein renounced the "vacuum tension" concept in favor of the "Big Bang." It simply could not exist in an expanding universe.

Hubble's "Big Bang" expanding universe has been accepted as dogma and has become the standard cosmology. This was done despite the fact that its originating data required an age of the universe which was incompatible with all other measures of cosmological age. It has been accepted by destroying the other data-confirmed component of a factual cosmology, Einstein's "vacuum tension" constant, his "cosmological constant" which we now know establishes the unrecognized fourth quantum dimension.

The quantum curvature model of Hubble's redshift-to-distance data establishes no age of the universe. As distance between stellar objects increase, so increases the amount of "vacuum tension kinking linear distance into curvature." Redshift is caused by the variance between the curved distance which light must follow and the linear distance. The quantum curvature model does not establish an age of the universe, it establishes a "visible limit" to the universe.

The gradual conversion of Hubble's "Big Bang" theory into a belief system has converted cosmology to mythology. By the 1960s the tension between Hubble's short age of the universe and other scientific beliefs about cosmological age had introduced a "poisonous effect" to science.

A "poisoned science" may be defined as one which deliberately assaults the scientific method with deception. The practice is "poisonous" because the practitioners are aware of the deception. They abandon all commitment to establishing fact using scientific methodology. Instead, they use the appearance of science to establish or reinforce a collectively held belief which furthers a collective or personal interest but which may be contradicted by authentic scientific facts. They hide the facts in favor of the belief.

The first example and the most damaging was the Baade and Sandage assault upon the Shapley/Eddington Cepheid scale which was the factual foundation for Hubble's 1929 data table. Baade's claim that Shipley's larger "scaling" sample was biased towards "dimmer type II" stars, while

Hubble smaller sample was not, can only be deliberate deception. It has no foundation in statistical mathematics. Similarly, Sandage's revision of Hubble's age by claiming "brightest star" error had no relevance to Shipley's actual methodology. Both were deceptions designed for the single purpose of revising the "Big Bang" age to something more acceptable to the current "Big Bang" belief upon which careers were and are founded. Sandage's corruption of the Shapley/Eddington Cepheid Scale and his exo-data revision of Hubble's Constant would be rewarded with a record 6 prestigious science awards.

2.   FRAUDULENT SCIENCE:
The Case of Particle Physics

The third "poisonous stream" in science was the development of so-called "particle physics." Beginning in the 1960s, a speculative pseudo-science known as "particle physics" was built upon the transference of applied data from inter-atomic nuclear measurements to fraudulent measurements of alleged particle fractions by deceptive use of particle accelerators.

"By the mid-1960's, physicists realized that their previous understanding, where all matter is composed of the fundamental protons, neutrons, and electron, was insufficient to explain the myriad new particles being discovered. Gell-Mann's and Zweig's quark theory solved these problems. Over the last thirty years, the theory that is now called the Standard Model of particles and interactions has gradually grown and gained increasing acceptance with new evidence from new particle accelerators.

"[In 1964] Murray Gell-Mann and George Zweig tentatively put forth the idea of quarks. They suggested that mesons and baryons are composites of three quarks or antiquarks, called up, down, or strange (u, d, s) with spin 0.5 and electric charges 2/3, -1/3, -1/3, respectively (it turns out that this theory is not completely accurate). Since the charges had never been observed, the introduction of quarks was treated more as a mathematical explanation of flavor patterns of particle masses than as a postulate of actual physical object. Later theoretical and experimental developments allow us to now regard the quarks as real physical objects, even though they cannot be isolated."[28]

The Gell-Mann/Zweig "quark theory" was "exo-data" in the sense that

---

28 *"Modern View (Standard Model) timeline: 1964 - present"*
http://www.particleadventure.org/other/history/smt.html

the alleged sub--particles had not been observed within or exiting the nucleus. They could be inferred because other "non constituent" particles had been allegedly detected. These were primarily "positrons" and "muons;" particles which did not directly compose atomic structure as did electrons, protons and neutrons . Gell-Mann and Zweig built an elaborate theory *" to explain the myriad new particles being discovered."*

The "exo-data" (outside data) model proposed that the known nuclear particles, the proton and the neutron, were composed of three quarks of different *"flavors"* and were lumped into a new category called "hadrons." These "hadrons" are then sub-categorized into "baryons" allegedly composed of three quarks (comprising protons and neutrons) and "mesons" allegedly composed of one quark and one alleged "anti-quark."[29]

The "mesons" were proposed to explain the well known process of beta decay by which protons decay to neutrons and neutrons decay back to protons. A specific "meson" imagined by Gell-Mann/Zweig was titled a "pion" which was used to explain the beta decay of a neutron back to a proton.

> *"The fraudulent nature of 'particle physics' is revealed by its explanation of neutron to proton conversions, an explanation which is incompatible with actual beta-decay empirical observations. It is claimed that both neutrons and protons are 'hadron-baryons' in that they both contain three quarks. The difference between neutrons and protons is in the type of quark. Neutrons contain one 'up' quark and two 'down' quarks. Protons contain one 'down' quark and two 'up' quarks.....*

> *"The conversion of the neutron to the proton is explained by a theoretical particle (the 'pion') which has been constructed wholly out of the quark model .... The 'pion' allegedly contains an 'up-quark' and a 'down anti-quark.' The 'pion' supposedly converts the neutron to a proton by quark matter/antimatter annihilation coupled with an*

29 *"Neutron decay rates decrease under the influence of a magnetic field proving the neutron's 'magnetic current' function (Scientific American article misunderstands its reported data),"* p.p. 3-4, Dawson, L. The Snake River N-Radiation Lab. http://paradigmphysics.com/neutron_decay.pdf

exchange with another form of quark. The pion's 'down anti-quark' allegedly annihilates one of the neutron's 'down quarks' and replaces it with the 'up quark' supposedly contained in the 'pion.' It is thus said that the two 'down' quarks and one 'up' quark of the neutron have become the two 'up' quarks and one 'down' quark of the proton."[30]

Not only does the "quark" model use an unproven particle, the "anti-quark" to explain the conversion of protons to neutrons, but the model cannot account for the masses of protons and neutrons at all.

"The accelerator-based "quark model" cannot account for neutron and proton masses as varying by an exact multiple of a merged electron's mass. In fact, the accelerator data have not and cannot, at all, account for the masses of "hadron-baryons" (protons and neutrons). Original 'quark theory' proposed that 'baryons' were composed of variations in three 'up' and 'down' quarks. The theory predicted these quarks must have a mass of approximately one-third the mass of the proton. However, subsequent measurements of quark masses ...... were much lower than predicted. 'Up quark' mass is currently assigned as between 3.9139 and 2.0163 times the mass of the electron. 'Down quark' mass is currently assigned as between 3.78344 and 2.6745 times the mass of the electron.[31] The mass of the proton as a multiple of electron mass is '1836.15267' times. The mass of the neutron is this value plus '2.525' times the mass of the electron. Measured quark masses......simply cannot account for measured neutron/proton masses. The 'quark model' can account for neutron/proton masses which are only 9-12 times the mass of the electron, not the actual '1836.15267' and '1838.67767' times."[32]

How can alleged "measures" of quark masses be so much lower than required if neutrons and protons were composed of quarks to make "hadron-baryons?" The answer is, I am afraid, that quark measurement has been deliberate fraud. The measure of quark masse was based upon a fraudulent use of linear electron accelerators.

---

30 Ibid. P.p. 3-4.
31 Calculated from reported mass values given as electron volts in *"Quarks"* at Georgia State University Hyperphysics.
    http://hyperphysics.phy-astr.gsu.edu/hbase/particles/quark.htm
32 *"Neutron decay rates decrease under the influence of a magnetic field .."* p.6. Op.Cit.

The quark model was initially proposed in 1964 which initiated a series of "exo-data" theoretical papers expanding the model even further. The first alleged experimental evidence for the actual existence of the quark wasn't provided until 1968-69.

> *"[1968-69] At the Stanford Linear Accelerator, in an experiment in which electrons are scattered off protons, the electrons appear to be bouncing off small hard cores inside the proton. James Bjorken and Richard Feynman analyze this data in terms of a model of constituent particles inside the proton (they didn't use the name 'quark' for the constituents, even though this experiment provided evidence for quarks.)"*

The alleged evidence of electrons *"scattered off protons,"* using linear accelerators, is then applied to the measure of quark masses. In a table of mass values showing the mass insufficiencies of the "up and down quarks," Georgia State University makes the following comment:

> *"The masses* [in the table] *should not be taken too seriously, because the confinement of quarks implies that we cannot isolate them to measure their masses in a direct way. The masses must be implied indirectly from scattering experiments* [electrons off protons]. *"*[33]

In 1968, Bjorken/Feynman were claiming they were "bouncing" (scattering) electrons off protons. However, it had been known for 40 years that accelerated electrons could not penetrate electron orbital distances to reach the nucleus. A free electron does not "collide" with a proton. The proton constitutes a hydrogen ion and the free electron enters an orbit around the proton/nucleus to produce a hydrogen atom. The velocity of the electron determines the distance of the orbit. High velocity electrons, as produced in linear accelerators (fractions of the speed of light), are scattered from the nucleus at x-ray orbital distances.

The electrons are thus "scattered," with a reduced exit velocities by releasing an x-ray burst. In 1925, Davisson-Germer[34] had shown that free

---

33 *"Quarks"* at Georgia State University Hyperphysics.
   http://hyperphysics.phy-astr.gsu.edu/hbase/particles/quark.htm
34 Davisson, C. J.; Germer, L. H. (1928-04-01). "Reflection of Electrons by a Crystal of Nickel". Proceedings of the National Academy of Sciences of the United States of

electrons shot at a nickel plate "scatter" at similar angles to the angles at which x-ray scatters off the plate. The angles of scatter are determined by the velocity of the electron and the wavelength of the x-ray.

> *"In 1925, The Davisson-Germer experiment determined that a beam of electrons fired against a nickel plate diffracted at angles similar to those of x-rays. X-rays are known to follow Braggs Law for diffraction when striking some crystalline structures.......The electron beam scattered off the nickel plate in a 'wavelike manner' in that the angle of diffraction was determined by the angle of incidence, length 0f crystallin bond and a proposed x-ray wavelength."[35]*

The interaction of an external electron moving at velocity towards a proton target (nucleus) is determined by atomic orbital structure. Specifically, the kinetic energy of velocity equals the energy of the sympathetic wavelength of an orbital position into which the electron's velocity allows it to be integrated. The electron "scatters" from this orbit.

The conventional shell/subshell structure cannot be penetrated by high-velocity electrons. Conventional orbitals contains a vibrational capacity which outputs light by accelerating and decelerating the electrons. When orbital velocity plus vibrational acceleration velocity exceeds the speed of light, the electron must fall out of the four-dimensional shell/subshell orbital structure and acquire a three dimensional orbit. Light output converts from *quantum harmonics* to conventional *Euclidean harmonics.* [36]

These non-quantum orbits are sympathetic with x-ray frequencies and do not allow the electron to be captured by the atom (under normal circumstances). The electron will be "scattered" from the atom, giving off a burst of x-ray at orbital frequency. The sympathy between x-ray burst and electron velocity is the reason *Davisson-Germer* detected equality between high-velocity-electron angles of diffraction and those of specific x-ray wavelengths.

The claims made in 1968 by James Bjorken and Richard Feynman that

---

America 14 (4): 317–322. Bibcode:1928PNAS...14..317D.
doi:10.1073/pnas.14.4.317. ISSN 0027-8424. PMC 1085484. PMID 16587341.
35 *Four Dimensional Atomic Structure*, p.p. 182. Dawson, L. Op. Cit.
36 Ibid. p.p. 16-39.

they were colliding electrons with protons to show that *"the electrons appear to be bouncing off small hard cores inside the proton"* was pure deception. For forty years science had demonstrated that atoms had scattered accelerated free electrons at x-ray-emitting orbital distances. An ignorant public was being sold the deceptions to further the career ambitions of men with speculative theories about nuclear protons being composed of smaller particles. These theories were formed outside the knowledge of actual atomic process.

Instead of being scattered by *"bouncing off"* protons, accelerated electrons were scattered by interactions with the *"quantum-squared field[37]"* which establishes the electron orbital positions within an atom. Because of their high velocities, accelerated, free electrons are assigned high energy, x-ray-sympathetic orbits which cannot capture the electron. The electron whips around the nucleus in this orbit and is ejected at a scatter angle which can be consistent with the scatter angle of the "rebroadcast" of an x-ray wavelength which may be impeded by the orbital. The accelerated electrons are not scattered by collision with the nuclear proton, but by the orbit which the electron tries to acquire.

All empirical data concerning electron scattering off the nucleus, prior to 1968, identified x-ray-associated orbital scattering. However, Bjorken/Feynman disguised this fact— asserting a direct scattering off the proton— in order to establish the model of the proton as composed of smaller sub-particles (quarks). Even as Bjorken/Feynman were conducting their deceptive experiments, linear electron accelerators in other parts of Stanford University where were being used experimentally to scatter electron streams off atomic targets to generate x-ray. These experiments (by Henry Kaplan and others) would ultimately lead to *Intensity-modulated radiation therapy (IMRT),* one of medical science's more successful forms of cancer treatment.[38]

The Bjorken/Feynman deception concerning atomic scattering by linear electron accelerators was adapted by the rest of science in order to

---

37 Ibid.. p.p. 16-70

38 *"History of Radiation Therapy"* UC San Diego School of Medicine: Dept. of Radiation Medicine and Applied Science.
http://healthsciences.ucsd.edu/som/radiation-medicine/about/Pages/history-radiation-therapy.aspx

establish the "standard model" of particle composition. The deception was carried forward despite the fact that the masses of quarks, as identified by the Bjorken/Feynman deceptive scattering measurements, were much to small when compared to the empirical measurements of proton and neutron masses.

Despite the lack of correct empirical confirmation, the whole "quark" system of sub-nuclear particles became the "standard model" of the nucleus. This "standard model" was built by deceptive interpretations of linear accelerator data. The particles assigned to the "hadron/baryon and hadron/meson" systems were never demonstrated by knowledgable accelerator data, but were established as "fact" by the "consensus" acceptance of the Bjorken/Feynman deception.

By the 1960s three actual sub-atomic particles had been confirmed and a fourth was about to be. A rather incomplete understanding of the atomic functions of the neutron, proton and electron had led to atomic fission energy, advances in our chemical bonding knowledge, the discovery of magnetic-fields/radio-wave interactions with the nucleus (nuclear magnetic resonance) as well as our understanding of the organic molecules underpinning life, to name but a few. A fourth particle which had long been theoretically predicted— the neutrino— was about to be empirically confirmed.

This fourth particle, which was first observed in 1968, is different from the others in that it is an "interactive" particle which is without mass when outside the atom (a vacuum particle). The revision of primitive quantum mechanics by quantum-dimensional mathematics would show how the *"proto-neutrino"* functions to bind the nucleus when integrated into the neutron. The *proto-neutrino* is a constituent part of the nuclear neutron. It functions to conduct a magnetic current down neutron-proton chains within the nucleus. This magnetic current, predicted by Pierre Curie[39], generates an electromagnetic field between the nucleus and the electrons; a field which allows the exchange of energy between the nucleus and the electron orbitals..

The *proto-neutrino* attached to the neutron is converted from an

---

39 *"The Curie-Quantum Nuclear Model and its Application to the Periodic Table of Elements."* in *Four Dimensional Atomic Structure,* p. 94. Dawson, L. Op. Cit.

electron which has been merged with a proton to produce the neutron. The conversion increases the mass of the merged electron by approximately "2.5 *times*."[40] This increase in mass is accomplished by an unrecognized form of quantum energy. The quantum-dimensional nuclear model identifies a new nuclear energy source which can be exploited[41]. However, recognition of the neutrino, its function and its energy potential cannot be disseminated to the scientific community because it has been eliminated as a possibility by the fraudulent particle physics established as the "standard model" in the 1960s.

The "standard model" is now taught as "fact" in all universities. The known and empirically-confirmed particles, the electron, proton and neutron, are artificially categorized as components of an invented sub-particle system. The proton and neutron are said to be composed of "quarks" and thus reside in the artificial category of "hadrons/baryons." They are said to be composed by three quarks, regardless of the fact that 40 years of the measurements of alleged quark "masses," using the LINAC Bjorken/Feynman deception, could only account for "0.65-0.4%" of required quark masses. *The 'quark model' can account for neutron/proton masses which are only 9-12* times *the mass of the electron* [42]*"* not the " *1836.15267 and 1838.67767 times"* electron mass of thee actual proton and neutron.

Included in the "standard model," now taught in all universities as "fact," is another subcategory of "hadrons" composed of particles which have never been observed and which incorporate a form of alleged "anti-matter." "Anti-matter" has never been proved to exist.

The "hadrons/mesons" are allegedly composed of a "quark" and an "anti-quark." "Mesons" are taught and accepted as "fact" without any research basis. The supposed "pion meson," a particle without any empirical foundation, is said to cause neutron-to-proton beta decay.

The Snake River N-Radiation Lab demonstrated an alternative model

---

40 *Four Dimensional Atomic Structure*, p.p. 78-81. Dawson, L. Op. Cit.
41 VIDEO: *"The SRNRL data proving the possibility of a nuclear generator using field controlled beta decay."*
   https://www.youtube.com/watch?v=i3_l9BJPJOQ&feature=youtu.be
42 *"Neutron decay rates decrease under the influence of a magnetic field .."* p.6. Op.Cit.

of neutron-to-proton beta decay by the suppression of *proto-neutrino* reconversion back to an electron using an externally applied electrical capacitance field. Control of beta-decay (*proto-neutrino* reconversion) using an electromagnetic field had never before been demonstrated, but it was able to establish the binding energy of the *proto-neutrino* as "0.511 MeV."[43]

The research results could not be reported in the scientific press because it proposed a cause-and -effect relationship which is outside the "standard model" of neutron-proton conversion. The point was driven home when the author attempted to present the SRNRL beta suppression data during an internet forum discussing neutrinos which was conducted by a University of Florida faculty member. The author was immediately mocked as a "crackpot" which included the posting of a picture of someone wearing a "tinfoil hat." The author was permanently excluded from the forum after asking for any actual data supporting the "standard model" neutrino theory which was being discussed.

The problem with the "quark" formulation is that it is contradicted by 50 years of actual nuclear research by some of the brightest minds in physics. Specifically, the neutron was known to be composed of an electron merged with a proton and not three speculative *"flavored quarks."*

In 1920, the father of nuclear science, Cambridge's Ernst Rutherford, proposed that the nucleus contained charge-neutral particles as well as the positively-charged protons which he had discovered. This neutral particle he named the "neutron."

Rutherford's postulation of the neutron was supported by the fact that the atomic weights of most elements are greater than can be accounted for by charge-bearing protons. The larger number of mass-bearing nuclear particles has become known as an element's "atomic mass" (identifying isotopes) and the number of charge-bearing protons as the "atomic number" which identifies the element's position within the Periodic Table.

---

43 *"Th-234 Beta Decay Ionization using an External Capacitance Field Demonstrates Electron/Nuclear Capacitance"* in *Four Dimensional Atomic Structure,* p.p. 49-59 Dawson, L. The Paradigm Company, 2015. ISBN 978-0941995351

Rutherford at first proposed that his neutron was composed of an intact electron orbiting a proton within the nucleus. However, it was rapidly determined that an intact electron could not exist within the nucleus. An intact electron confined to the nucleus would have greater energy than that binding it within the nucleus. It could easily escape the nucleus.

> *".... an electron confined to a region the size of an atomic nucleus has an expected kinetic energy of 10–100 MeV. [see footnotes in original article] This energy is larger than the binding energy of nucleons and larger than the observed energy of beta particles emitted from the nucleus"* [44]

The problem of the excessive amount of energy provided intact electrons confined to the space of the nucleus is resolved by the *"merging"* of the electron with the proton to produce the neutron.

The *Snake River N-Radiation Lab* has experimentally proved the binding energy which keeps the *merged* electron contained within the neutron. Our lab was able to control the beta decay of Thorium 234 neutrons using an externally applied capacitance field. [45] When the beta-decaying electron voltage was suppressed to below 0.511 MeV, the ejection of the beta electron from the neutron (*proto-neutrino* reconversion) was fully suppressed and the conversion of the neutron to a proton prevented.

The binding energy supplied by the nucleus to retain an electron *merged* with the neutron is only "0.511 MeV." When the neutron is removed from the nucleus and the binding energy is no longer supplied, the neutron rapidly decays back to a proton. [46]

For an electron *merged* with the neutron, the amount of nuclear binding energy required to retain the electron is only "0.511 MeV." For an intact electron contained within the nucleus as a separate particle, the

44 *"Problems of the nuclear electrons hypothesis"* in *Discovery of the Neutron* Wikipedia.   https://en.wikipedia.org/wiki/Discovery_of_the_neutron
45 *"Th-234 Beta Decay Ionization using an External Capacitance Field Demonstrates Electron/Nuclear Capacitance"* in *Four Dimensional Atomic Structure*, p.p. 49-59 Dawson, L. The Paradigm Company, 2015. ISBN 978-0941995351
46 *"Neutron decay rates decrease under the influence of a magnetic field proving the neutron's 'magnetic current' function"* Dawson, L Op. Cit.

amount of binding energy was calculated as requiring *"10-100 MeV."* Rutherford's original "duplex" model of the neutron (electron orbiting a proton within the nucleus) is simply not possible and has been replaced by the model of the neutron as composed by an electron which has been merged with a proton.

By 1934, the concept of the electron being merged with the proton to produce a *"unitary"* neutron rather than Rutherford's original *"duplex"* concept had been proved.

James Chadwick, Rutherford's assistant in Cambridge's Cavendish Laboratory, had previously proved the existence of the neutron. He had bombarded paraffin with alpha particles (helium nuclei) to observe an unknown "neutral emission" (one unaffected by a magnetic field). He then observed the impact of these "neutral emissions" upon various gases. Such"neutral emissions" had been previously proposed to be gamma radiation. However Chadwick showed that the neutral emissions didn't ionize gases, as did gamma radiation, and, therefore, they must be the neutral particle predicted by Rutherford, the neutron. .

Chadwick also proposed a test for the neutron to determine if they were composed by the merging of the electron with the proton (unitary particle) or were a "proto hydrogen atom" (duplex particle) contained within the nucleus.

> *"As posed by Chadwick in his Bakerian Lecture in 1933, the primary question was the mass of the neutron relative to the proton. If the neutron's mass was less than the combined masses of a proton and an electron (1.0078 u), then the neutron could be a proton-electron composite..... If greater than the combined masses, then the neutron was elementary like the proton.[47]" [48]*

In 1935, the year Chadwick received the Nobel Prize for his measurement of the neutron, he made the first precise measurement of the

---

47 *Chadwick, J. (1933). "Bakerian Lecture - The Neutron". Proc. Roy. Soc. 142 (846): 1–25. Bibcode:1933RSPSA.142....1C. doi:10.1098/rspa.1933.0152. [FOOTNOTE IN ORIGINAL]*

48 *"Proton–neutron model of the nucleus"* in *Discovery of the neutron .* Wikipedia
https://en.wikipedia.org/wiki/Discovery_of_the_neutron

mass of the neutron which proved it was *"elementary"* (merged electron) rather than *"composite"* (a nuclear proton orbited by an electron).

> *"The issue was resolved in 1935 when Chadwick and his doctoral student Maurice Goldhaber, reported the first accurate measurement of the mass of the neutron. They used the 2.6 MeV gamma rays of Thallium-208 (^{208}Tl) (then known as thorium C) to photodisintegrate deuterium….*
>
> *"Chadwick and Goldhaber found the neutron's mass to be slightly greater than the mass of the proton (1.0084 <u>u</u> or 1.0090 <u>u</u>, depending on precise values used for the proton and deuteron masses), [i.e. greater than 1.0078u]….The mass of the neutron was too large to be a proton-electron composite, and the neutron was therefore an elementary particle."* [49]

The mass of the neutron (at 1.00866u or 1.67493e -27 kg) is currently known to equal the mass of the proton *plus* approximately 2.5 *times* the mass of a merged electron.[50] Chadwick's relatively precise 1935 neutron measurement in comparison with the mass of the proton would have allowed science to recognize that the merged electron had increased in mass by approximately 2.5 times.

The mathematics which would explain Chadwick's *"2.5 times"* gain in mass for the merged electron did not exist in 1935. The replacement of 1935 primitive quantum mechanics with quantum-dimensional mathematics would not occur until 2008.[51] Chadwick could not have recognized that the bond between the electron and proton was the quantum force which sustained a fourth, unrecognized quantum geometric dimension; the same force which curved space and which better explained Hubble's 1929 redshift to stellar distance data by which Hubble had incorrectly postulated his "Big Bang" theory.

Chadwick would not have recognized that the bond between the

---

49 Ibid.
50 *"The mass of the neutron can be calculated by the quantum geometric model,"* in *Four Dimensional Atomic Structure*, p.p. 78-80. Dawson, L. Op. Cit.
51 *"The Mystical Quantum vs Quantum-Dimensional Math"* in *Four Dimensional Atomic Structure*, p.p. 8-15. Dawson, L. Op. Cit.

orbiting electron and the proton was the *"quantum squared"* or that, when the electron was merged with the proton, the bond must be inverted to become the *"inverted quantum squared."* He would not have recognized that quantum force must, geometrically, increase the mass of the electron by 2.5 *times* during the conversion of the bond from the quantum squared to the inverted quantum squared.[52]

Nonetheless, Chadwick had kept faith with the strict scientific methods of measurement by which the Rutherford Group had built the modern model of the atom. Rutherford had been the first to describe the orbital model of atomic electrons; to demonstrate the existence of the proton and to predicte the existence of the neutron. Chadwick, Rutherford's assistant in the Cavendish Labs, had proved Rutherford's predicted "neutron" by demonstrating that the charge-neutral radiation emitted by alpha-particle bombardment of paraffin could not be gamma radiation because it did not ionize gases as gamma was known to do.

Further, Chadwick had devised a test to see if the neutron were a proton being orbited by an electron within the nucleus, or a unitary particle made by the merging of an electron with the proton. He had used deuterium hydrogen to make the first modern measurement of the neutron's mass. The deuterium nucleus is composed of one proton and one neutron.

The difference between the mass of the deuterium nucleus and a standard hydrogen nucleus is the mass of a neutron. Chadwick theorized that, if the neutron's mass is less than the mass of the proton *plus* the mass of the electron, then the neutron would be a proton orbited by an electron. If the neutron's mass were greater than the mass of the proton *plus* the mass of an electron, then the neutron is a unitary particle composed by the merging of the electron with the proton.

Chadwick measured the neutron's mass as equal to the mass of the proton *plus* between "2.05 *times* and 3.14 *times*" the mass of the electron (current quantum-dimensional value "2.525 *times*" electron mass). Thus, Chadwick proved the neutron was a unitary particle composed of an electron merged with a proton.

---

52 Ibid. *"The Quantum Geometric Model of the Neutron,"* p.p. 71-93.

The Desertion of Chadwick's Neutron-Mass Data in favor of Rudimentary Particle Physics

James Chadwick's relatively accurate measure of neutron mass was ignored in a mad rush to explain beta decay with rudimentary "anti-matter" particle physics. Beta decay is the radioactive process by which a nuclear proton is converted to a neutron or a nuclear neutron is converted to a proton. These conversions transform an element into a different element. Radioactive beta decay is one of the primary processes by which elements are transitioned. For example the radioactive chain for Uranium 238 incorporates a stage in which transitional Thorium 234 double beta decays to create Uranium 234[53] ultimately leading to the end of the decay chain which is lead.

In 1934, the year before Chadwick made his definitive measurement of the neutron's mass, Frédéric and Irène Joliot-Curie announced they had discovered that the conversion of a proton to a neutron was accompanied by the emission of an proposed "anti-particle" called a "positron." A "positron" is allegedly the "anti-matter" version of an electron. The "positron" has the same mass of the electron but an opposite charge. The charge of the electron is negative, while the charge of the "positron" is positive. The particles allegedly attract one another and supposedly "annihilate" one another in a high-energy burst.

Joliot-Curie had bombarded aluminum with alpha particles to produce "Phosphorus-30'" which decayed to "Silicon-30" by converting a proton in the phosphorus nucleus to a neutron (positive beta decay). [54] Using a cloud chamber, the Joliot-Curie team found that the positive beta decay was accompanied by the emission of a positively charged particle with a mass much less than than that of the positively charged proton. The charge-to-mass ratio could be determined by the amount and direction of curvature detected in the cloud chamber under a magnetic field.

The Joliot-Curies concluded that the conversion of a proton to a neutron was caused by the ejection of a "positron." This characterization

---

53 *"Th-234 Beta Decay Ionization ......"* in *Four Dimensional Atomic Structure,* p.p. 49-59 . Dawson, L. Op. Cit.

54 **Research Profile:** Irène Curie Nobel Prize in Chemistry 1935 *"in recognition of their synthesis of new radioactive elements".* p.p. 3. Lindau Nobel Laureate Meetings No. 66. *by Luisa Bonolis..*
    http://www.mediatheque.lindau-nobel.org/research-profile/laureate-joliot-curie

was made without rigorous measurement of mass. It was partially "exo data," and accepted to further the concept of "anti-particles."

The problem with the Joliot-Curies' "positron ejection" beta decay is that it is incompatible with the actual Chadwick measurements of the masses of nuclear particles. How could the proton eject a particle of the mass of the electron and arrive at a new mass which was equal to the original mass of the proton *plus* 2.5 *times* the mass of the electron.

A further problem for the Joliot-Curie model of positive beta decay occurred when an alternative method of proton-to-neutron conversion was discovered. Nuclear protons could be converted to neutrons by capturing one of the atom's electrons (K-electron capture).

> *"The theory of electron capture was first discussed by Gian-Carlo Wick in a 1934 paper.....K-electron capture was first observed by Luis Alvarez, in vanadium-48. He reported it in a 1937 paper in Physical Review[55]."[56]*

In violation of scientific standards, multiple causes of proton-to-neutron conversion were accepted. A "high energy" conversion was said to be caused by "positron" emission. A "low energy" conversion was supposedly by electron capture. The "first cause" also violated the Law of the Conversion of Matter. A lighter particle ejects a fraction and is said to become heavier. This violation of all logic was accepted because it was believed that the proposed theoretical "anti-matter" particle operated in mysterious ways which excluded it from the rules governing the conservation of matter. Such irrational beliefs were the legacy of "primitive quantum mechanics." [57]

The "anti-matter positron" was based upon a set of observations made by Cal Tech's Carl Anderson using the Mount Wilson Observatory cloud chamber (published 1933). Out of 1300 photographs of cosmic ray tracks made in the Mt. Wilson cloud chamber under a magnetic field, Anderson

---

55 Alvarez, Luis W. (1937). "Nuclear K Electron Capture". Physical Review 52: 134–135. Bibcode:1937PhRv...52..134A. doi:10.1103/PhysRev.52.134.
56 Electron capture-History. Wikipedia.
   https://en.wikipedia.org/wiki/Electron_capture#History
57 *"The Mystical Quantum vs. Quantum-Dimensional Math"* in *Four Dimensional Atomic Structure*, p.p. 8. Dawson, L. Op. Cit.

discovered 15 which had been bent in the direction of a positive charge but which had a charge-to-mass ratio less than that of a positively charged proton. A charged particle in a field-influenced cloud chamber leaves a trail which is bent by the field and the amount of curvature is determined by the charge and the mass of the particle. With greater charge, the curvature is greater. With greater mass, the curvature is less.

Anderson concluded that the particles were "positively charged electrons" (positrons) which contained a positive "elementary charge" (1.602176565e-19 coulombs) equal to that of the electron and proton but with a mass equal to that of the electron (9.10938291e-31 kg). His data, in fact, did not support this conclusion. There was nothing of the precise rigor which James Chadwick had applied to his measurement of the mass of the neutron.

Anderson's data was ambiguous and may have implied a particle other than the "positron," albeit one which awaited future discovery. Anderson's data might have suggested a possible "energy to mass conversion" which could result from the process by which a merged electron is converted to the *proto-neutrino* during positive beta decay.

Specifically, quantum-dimensional mathematics accurately calculates the energy gain recorded when a merged electron is converted to the *proto-neutrino* during positive beta-decay.[58] The excess energy of the conversion is "0.873 MeV." This energy excess allows for the energy-creation and emission of a reduced-density, positively charged particle which is a mirror of the "2.525" *proto-neutrino* mass increase. This ejected "mirror" particle would have a reduced charge of "2/3" of a proton's charge. This proposed "mirror image" of the merged electron's mass increase during beta decay would fit the actual Anderson data.

> *"If these particles carry unit positive charge the curvatures and ionizations produced require the mass to be less than twenty times the electron mass. These particles will be called positrons. Because they occur in groups associated with other tracks it is concluded that they*

---

58 *"A Quantum-Dimensional Model of Positive Beta Decay is revealed by Medical Science's "PET Scan" (New Model Accurately Calculates Reported Gamma Emission Energy)"* Dawson, L. SRNRL.
http://paradigmphysics.com/PET-scan-pos-beta-decay-theory2.pdf

must be secondary particles ejected from atomic nuclei."[59]

Anderson's cloud-chamber tracks may have actually recorded *"proton ghosts"* created and ejected during positive beta decay by the excess energy of *electron-to-proto-neutrino* conversion within the nucleus. The amount of track curvature would have calculated to a particle with "2.525 *times*" the mass of the electron, but with the charge of the proton; a "mirror image" of the "2.525 *times*" mass increase experienced by the merged electron during positive beta decay. Anderson's positively charged particles had masses which ranged below *"twenty times the electron mass."*

We can't let Anderson off the hook by simply noting he might have misunderstood his data. This "anti-electron" proposed by Anderson was never held as a hypothesis which needed to be tested and proved experimentally. Despite the fact that the Anderson data was inconclusive, relative to the mass of an alleged "positron," the "anti-particle" model was accepted because it fit the "exo-data theorizing" from primitive quantum mechanics, especially that of Cambridge's Paul Dirac.

> *"Dirac published a paper in 1931 that predicted the existence of an as-yet unobserved particle that he called an 'anti-electron' that would have the same mass as an electron and that would mutually annihilate upon contact with an electron.*[60]" [61]

Despite these recorded variations in mass for the Anderson data, the explanation offered is of a positively charged particle of a single mass which is equivalent with the electron mass. The "positron" explanation of the data was accepted "exo-data" because Cambridge University's Paul Dirac had predicted the existence of an "anti-electron" two years earlier. He had theoretically proposed anti-matter particles in support of a complex equation (the Dirac equation) which argued all particles had a "negative energy state" and which he had advanced to explain the

59 *"The Positive Electron."* Anderson, Carl D. (California Institute of Technology, Pasadena, California) Physical Review, vol. 43, Issue 6, pp. 491-494. Date 03/1933
60 P. A. M. Dirac (1931). "Quantised Singularities in the Quantum Field". Proc. R. Soc. Lond. A 133 (821): 2–3. Bibcode:1931RSPSA.133...60D. doi:10.1098/rspa.1931.0130.
61 *"Positron"* Wikipedia.   https://en.wikipedia.org/wiki/Positron

anomalous "Zeeman Effect."[62]

The "Zeeman Effect" is the demonstration by Pieter Zeeman that a set of naturally occurring light doublets in the sodium spectrum could be multiplied by application of a magnetic field. The doublet consisted of two very close lines on the spectrograph with very narrow offset wavelengths. The Zeeman application of a magnetic field further split the naturally occurring two lines into ten lines.

Although conducted in 1896, the Zeeman experiment became a major problem for primitive quantum mechanics because all of quantum physics had been built upon the relationship of electron orbitals and associated light emissions. They had used Planck's Constant and the quantized (by the Rydberg Formula) spectrographic lines of hydrogen to identify the quantum orbits of electrons. The sodium doublets were a light emission associated with sodium electrons which were hard to explain using the primitive quantum mechanical orbital model. Zeeman's magnetic field expansion of the natural doublets were even more so.

The sodium "D lines" and their further splitting by Zeeman required increasingly esoteric and complex quantum mechanical equations to attempt an explanation. This mathematical reaching to explain the Zeeman Effect had led Dirac to his "negative energy state" and "anti-electron."

In point of fact, the explanation for the sodium light doublet and Zeeman's further splitting of it has a rather mundane and somewhat commonplace mathematical explanation. Primitive quantum mechanics could not recognize the sodium "D lines" and their Zeeman modification because they had developed an incorrect infill pattern for the quantum electron orbitals. The correction by quantum-dimensional mathematics shows the "D lines" are caused by this corrected orbital electron-infill pattern and the Zeeman Effect by a mathematically predicted increase in orbital electron voltages from Zeeman's magnetic field. [63]

Using an incorrect model for electron orbitals, Dirac developed a set of equations which would explain the Zeeman Effect by postulating a duel energy state for atomic particles. Particles were said to have a "negative

---

62 *"Positron"* Wikipedia.   https://en.wikipedia.org/wiki/Positron.
63 *Four Dimensional Atomic Structure*, p.p. 65-70. Dawson, L. Op. Cit.

energy state" as well as a "positive energy state;" states which would produce a high-energy "annihilation" reaction when brought in contact with one another.

Originally, Dirac had postulated that the positively charged proton was the negative energy state for the negatively charged electron. However, this solution was rapidly opposed by nuclear scientists.

> *"Robert Oppenheimer argued strongly against the proton being the negative-energy electron solution to Dirac's equation. He asserted that if it were, the hydrogen atom would rapidly self-destruct. Persuaded by Oppenheimer's argument, Dirac published a paper in 1931 that predicted the existence of an as-yet unobserved particle that he called an "anti-electron" that would have the same mass as an electron and that would mutually annihilate upon contact with an electron."*[64]

In 1931, incorrect quantum mechanical theory required the existence of an "anti-electron" or positively charged electron which could annihilate a normal electron in order to explain the Zeeman Effect. Two years later, Carl Anderson measured positively charged particles with a fraction of the mass of protons using the Mount Wilson Observatory cloud chamber. Those measurements were identified as a previously theorized "anti-electron" proposed by primitive quantum mechanics. This "fit" was made despite the fact that the actual Mount Wilson data contradicted the physical requirements which the theorist, Paul Dirac, had placed upon his "anti-electron."

The most important factor identifying the actual particles recorded in the Mount Wilson cloud chamber is the lack of "annihilation" data. The Dirac "anti-electron" is supposed to "annihilate" itself and an electron on contact. This "annihilation" is supposed to release high energy gamma with "1.022 MeV" of energy[65]. However, none of Anderson's "positron" tracks end in the diffused cloud pattern characteristic of gamma bursts. Cambridge's C.T.R. Wilson, the inventor of the cloud chamber, described

---

64 *"Positron"* Wikipedia. Op. Cit.
65 *"A Quantum-Dimensional Model of Positive Beta Decay is revealed by Medical Science's 'PET Scan'"* Dawson, L. SRNRL.
   http://paradigmphysics.com/PET-scan-pos-beta-decay-theory2.pdf

the characteristics of gamma as recorded in his cloud chamber.

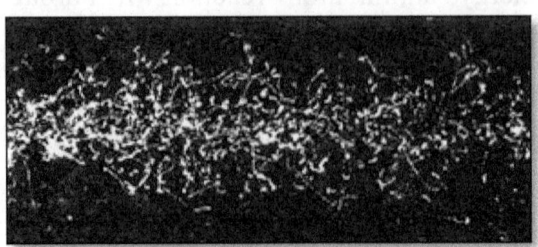

Photograph of a typical X-ray cloud.
Condensation on the ions appears in white.

"Wilson also tried using radium bromide to produce gamma-radiation. He saw the same cloud tracks as he had seen for beta-particles [electrons], but the tracks were in all directions, rather than directed from the source. The walls of the chamber were absorbing the gamma-radiation and emitting beta-radiation from all directions."[66]

Had Anderson's curved cloud-chamber tracks identified Dirac's "negative energy" electron? Or had they identified "proton ghosts" created by the proto-neutrino conversion of a merged electron during the beta decay of a proton to a neutron? Such particles would have a "2/3 positive charge" and a "single-charge" mass of 2.525 *times* the mass of an electron. It would not "annihilate" an electron in a gamma burst, but electron contact would decay the *"mirror proto-neutrino"* to a neutrino; a neutrino which would disappear from the cloud chamber record.[67]

If the tracks identified Dirac's "anti-electron," electron-positron annihilation should have occurred in the electron rich atmosphere of the ionized water vapor contained within the Mount Wilson cloud chamber. Yet the record shows none of the multidirectional cloud tracks which C.T.R Wilson had discovered to be characteristic of gamma emissions. None of Anderson's tracks ended in Dirac's predicted "annihilation" from his proposed "anti-electron." Yet Anderson still identified the tracks as being produced by Dirac's proposed "positron anti-electron."

"Mirrored Proto-Neutrino" **Emissions during Proton Beta Decay**
If the Mount Wilson data actually indicated cosmic proto-neutrino

---

66 *"The Cloud Chamber-Wilson, 1912: 13. Gamma Rays and X-ray"* Cavendish
    Laboratory, University of Cambridge
    http://www-outreach.phy.cam.ac.uk/camphy/cloudchamber/cloudchamber13_1.htm
67 Reduced in charge and density, the "ghost" could merge with an electron.

"ghost " emissions with no evidence of "anti-electron" annihilation, what about the Joliot-Curie "positron emissions" which were recorded when they initiated the decay of Phosphorus 30 (atomic number "15") to Silicon 30 (atomic number "14"). The decay was accomplished by converting a phosphorus proton to a neutron and thus reducing the atomic number from "15 to 14." [68]

Before recognizing the inadequacy of the Joliot-Curie analysis of their data, we must first discover what proton-to-neutron conversion has proven to be by modern mathematical science. The conversion of a proton to a neutron, or "positive beta decay," occurs when the proton-to-neutron ratios within the nucleus becomes unbalanced for odd numbered elements.

Chains of protons/neutrons within the nucleus conduct a Curie magnetic current. For the first twenty elements, known as the "flat nuclei," the nuclei of odd numbered elements must contain one more neutron than they contain protons. This provides a neutron-to-proton ratio of "(1+1):1."

The extra neutron for odd-numbered elements is necessary to split the charge of the odd numbered proton to balance two streams of magnetic current. If the magnetic current streams receive an unbalanced number of charges, an electron is "sucked" into the nucleus, converting a nuclear proton to a neutron.[69]

The Joliot-Curie beta decay of Phosphorus-30 exemplifies the required neutron-to-proton ratio for odd numbered elements. Phosphorus is an odd numbered element with an atomic number of "15." The stable isotope of Phosphorus is "Phosphorus-31" which provides a "1+1:1" neutron-to-proton ratio (16 neutrons to 15 protons).

The Joliot-Curie study used alpha particles (composed of a helium nuclei) to bombard Aluminum-27 (atomic number "13") to produce Phosphorus-30 (atomic number "15") plus one neutron.[70] Phosphorus-30

---

68 **Research Profile:** Irène Curie Nobel Prize in Chemistry 1935 *"in recognition of their synthesis of new radioactive elements".* p.p. 3. Lindau Nobel Laureate Meetings No. 66. *by Luisa Bonolis..*
 http://www.mediatheque.lindau-nobel.org/research-profile/laureate-joliot-curie
69 *A Quantum-Dimensional Model of Positive Beta Decay......."* Dawson, L. Op. Cit.
70 **Research Profile:** Irène Curie Nobel Prize in Chemistry 1935 *"in recognition of*

has an unbalanced neutron-to-proton ratio "1:1" (for an odd-numbered element). It positively beta decayed to even-numbered Silicon-30 because the two nuclear magnetic current streams within the Phosphorus-30 nucleus contained unequal numbers of freed charges.[71]

The Joliot-Curies proved the Phosphorus-30 which they had created was beta decaying to even numbered Silicon-30 (atomic number "14"). Using a cloud chamber under a magnetic field, similar to that used by Anderson, they had detected positively charged particles with a mass which was a fraction of proton masses; the same type of particles Anderson had detected in the Mount Wilson observatory. However, the Joliot-Curie particles were not cosmic radiation but were being emitted by the beta-decaying Phosphorus-30. From this they concluded that the conversion of a proton to a neutron occurred by the ejection of Dirac's "positron" anti-electron. However, the cloud-chamber again revealed no gamma radiation bursts which would indicate the predicted "annihilation."

Within the same time frame (1934) during which the Joliot-Curies had allegedly "proved" that lighter protons decay to heavier neutrons by emitting a "positron," the nuclear physicist, Enrico Fermi issued a theoretical paper which accurately anticipated the actual explanation of the Joliot-Curie data. Fermi proposed that beta decay could be explained as an exchange between electrons and a newly predicted particle the "neutrino."

> *"The Hamiltonian* [energy equation] *of the system consisting of heavy and lightweight particles must be chosen that each transition from a neutron to a proton is associated with the creation of an electron and a neutrino. The reverse process (change a proton into a neutron) must be associated with the annihilation of an electron and a neutrino. Note that by this, conservation of charge is assured."[72]*

A neutrino is a massless particle without charge which is identified as a "vacuum particle" in quantum-dimensional physics. It was first theoretically described by Wolfgang Pauli in 1931 and named by Erico

their synthesis of new radioactive elements". p.p. 3. Lindau Nobel Laureate Meetings No. 66. *by Luisa Bonolis..*
http://www.mediatheque.lindau-nobel.org/research-profile/laureate-joliot-curie#page=3
71 *A Quantum-Dimensional Model of Positive Beta Decay......."* Dawson, L. Op. Cit. .
72 "Fermi's Theory of Beta Decay, A complete English translation." Wilson, F.L. American Journal of Physics, Vol. 36, No. 12, December 1968, p.p .1151

Fermi in 1933 who described a possible energy function.

> *"[In 1933] Enrico Fermi proposes 'neutrino' as the name for Pauli's postulated particle. He formulates a quantitative theory of weak particle interactions in which the neutrino plays an integral part"[73]*

It wan't till 1968 that the actual neutrino was detected leaving Cherenkov light trails in water. Anything traveling faster than the restricted speed of light in water leaves a tell-tale light trail. The Cherenkov light trails were restricted to neutrinos because they are the only form of radiation which can penetrate deep into the earth. These "massless" neutrinos had passed through miles of rock.

> *"[1968] An experiment deep underground in the Homestake mine in South Dakota makes the first observation of neutrinos from the sun. But experimenters see far fewer neutrinos than solar models had predicted."[74]*

The *"integral part"* which Fermi postulated for the neutrino was a role in the beta decay conversions of neutrons and protons. He visualized the two forms of beta decay as produced by the creations and destructions of electrons and neutrinos. This represented an alternative explanation for positive beta decay (conversion of protons to neutrons) to that of the Joliot-Curies. The conversion of protons to neutrons was not caused by ejection of Dirac's "anti-electron" but by *"the annihilation of an electron and a neutrino."*

Fermi's *"annihilation,"* however, was not the same as proposed for the electron-positron annihilation:

> *"Electrons (or neutrinos) can be created or annihilated. This possibility, however, is not analogous to the creation or annihilation of an electron-positron pair."[75]*

In his concept of proton/neutron conversion, Fermi had anticipated the actual relationship which exists between the proton and the neutron. The proton is converted to a neutron when one of the electrons in the atomic

---

73 *"The Story of the Neutrino: Historic Events in Neutrino Physics"* FermiLab
   http://www.fnal.gov/pub/inquiring/physics/neutrino/discovery/history.html
74 Ibid.
75 "Fermi's Theory of Beta Decay, A complete English translation." p. 1151. Op. Cit.

orbit is merged with the proton and converted to a *"proto-neutrino."* By Fermi's definition , the merged electron has, in fact, been *"annihilated"* in the sense that it has been *"anti-created"* rather than being permanently removed from existence in an energy burst, as with theoretical *"electron-positron annihilation."*

Fermi had been correct in recognizing a role for electrons and neutrinos in the conversion of protons to neutrons. However, his theory was plagued by the inadequacies of *primitive quantum mechanics*. It would be 70 years before the irrationalities of *primitive quantum mechanics* would be resolved by the recognition of a fourth quantum dimension[76]; before *Quantum-Dimensional Mathematics* would provide an exact model for the neutron and its function within the nucleus.

This quantum-dimensional model is proved by the contemporary energy measurements by medical science of proton-to-neutron conversions.[77] A positive beta-decaying isotope of Fluorine is used for a medical imaging device which is known as a "PET scan" (Positron Emission Tomography). A radioactive isotope of dextrose sugar is taken into the cells which has replaced an oxygen atom with a beta-decaying Fluorine atom. The decay of a Fluorine-18 proton to a neutron converts the atom to Oxygen-18 and releases gamma radiation which is detected with a gamma camera. The Fluorine-18 atomic number is "9;" making it an uneven numbered element, but its neutron-to-proton ratio is "1:1" which creates a set of unequal magnetic current flows and which initiates the beta decay.[78]

The device is labeled *"Positron Emission"* since the Joliot-Curie model of proton-to-neutron conversion is being used. However, the actual energy measurements demonstrate that it is operating as a variety of the Fermi *"electron-neutrino"* model, not the *"positron-electron"* model. It is assumed that the beta-decay gamma radiation emissions which the camera records are the energy output of an "electron-positron" annihilation resulting from the release of a "positron" anti-electron during the decay.

---

76 *"The Mystical Quantum vs. Quantum-Dimensional Math"* in *Four Dimensional Atomic Structure*, p.p. 8=15.   Dawson, L. Op. Cit.

77 *"A Quantum-Dimensional Model of Positive Beta Decay is revealed by Medical Science's 'PET Scan' ,"* Dawson, L. SRNRL.
http://paradigmphysics.com/PET-scan-pos-beta-decay-theory2.pdf

78 Ibid.

It is assumed that the Fluorine-18 decay always produces such an annihilation resulting in a gamma emission— an event which was never observed by either Anderson's cloud chamber"positron" observations nor those of the Joliot-Curies from the positive beta decay of Phosphorus-30. In neither case did the "positron" tracks end in the tell-tale tracks of gamma emissions from *"annihilation."*

A comparison between the Joliot-Curies' "positron emitting" Phosphorus-30 beta decay and the gamma emitting Fluorine-18 beta decay used by medical science reveals how proton-to-neutron conversion actually works as well as what the alleged "positron" actually is. The energy released by medical science's Fluorine-18 decay is much better documented and can be calculated directly by the quantum-dimensional model of the neutron.[79]

Specifically, the excess energy which the decay process can output has been measured as "873.431± 0.593 keV " (approx. 0.873 MeV).[80] This excess decay energy can either be output as a particle or as gamma radiation, depending upon the nuclear configuration of the magnetic-current circuitry specific to the decaying element's nucleus.

The potential particle released, however, is not the theoretical "anti-electron" or "positron." It is a *"mirror proto-neutrino"* which can decay to a neutrino when merged by contact with an electron. This electron/*mirror-P-neutrino* merger is a Fermi *"electron-neutrino annihilation."* It is not an alleged Dirac *"electron-positron annihilation."* It does not release *annihilation energy*, but absorbs the energy to facilitate the electron-neutrino conversion.

Proton-to-neutron conversion occurs when unequal nuclear magnetic current streams flow into the charge-transmission proton. This "sucks" an orbiting electron into the proton for merger. The merged electron decays the proton to a neutron which leaves one current-circuit unbalanced with three neutrons in a row. If the element's nucleus is somewhat complex, as

79 *"A Quantum-Dimensional Model of Positive Beta Decay is revealed by Medical Science's 'PET Scan.'"*   Dawson, L. SRNRL.
      http://paradigmphysics.com/PET-scan-pos-beta-decay-theory2.pdf
80 *"Fluorine-18;"*   https://en.wikipedia.org/wiki/Fluorine-18

with Joliot-Curies' Phosphorous-30, It will require some time to rearrange the two excess chained neutrons into the daughter element's nuclear structure. The "0.873 MeV" energy of decay cannot be input back into the magnetic current while the nuclear structure is being rearranged. The energy of neutron conversion will be used to create the *mirror-P-neutrino* and eject it from the proton-*cum*-neutron.

The Joliot-Curies cloud chamber detected such *mirror-P-neutrino* emissions as the result of the decay of Phosphorous-30. The particles could not have been "positron" anti-electrons because they were not exhibiting gamma radiation from the predicted annihilation reaction. The cloud chamber detected no gamma radiation from the proton-to-neutron conversions, only the trails of positively-charged particles with a fraction of the mass of a proton. They detected only *mirror-P-neutrinos* which would disappear, without an energy burst, into massless neutrinos in a *Fermi annihilation* encounter with an electron.

As might be imagined, the mechanics governing Fermi *"electron-neutrino annihilation"* are somewhat complex. The essentials are the following: the electron is a partially quantum (vacuum) volume which only has mass when its fourth-dimensional charge is anchored to a Euclidean mass; the *mirror P-neutrino* (MPN) has a "2/3" positive charge which is anchored to only two of its three dimensions of volume; The MPN has an extremely low density, which is only 0.1375% of a proton's density and which is equal to the *proto-neutrino's* mass of 2.525 times the mass of an electron; the electron cannot gain mass by anchoring to the MPN because the MPN's own charge is disconnected from the third dimension of mass which the electron requires; on annihilation by an electron, the remnant of the MPN charge must be rotated 90° to the charge of the electron and spun; the low-density mass of the MPN is applied to the mass equivalence required to transform the electron into the encapsulating shell for the spinning MPN free charge and to convert the particle into a massless neutrino.

During positive beta decay, the conversion of the merged electron to the *proto-neutrino* increases the mass of the electron 2.525 *times*. The conversion to the *proto-neutrino* consists of rotating the charge of the proton relative to the charge of the electron and encapsulating the proton's charge in free space within a shell composed by the merged electron. The *proto-neutrino* is attached to the surface of the neutron and is a component of its mass. The spin of this free charge becomes the impeller for a Curie magnetic current composed of freed proton charges which are conducted down proton-neutron chains towards an end-chain transmission proton.[81]

For the first twenty elements, proton-to-neutron conversion occurs for isotopes of odd-numbered elements which do not provide one more neutron than protons in their nuclei. A neutron-to-proton ratio of "1:1" assures that an unequal number of charges will be contained within the magnetic current circuits and will initiate positive beta decay[82].

Any conversion of a proton to a neutron requires a realignment of the nuclear neutron/proton chains. If the nucleus of the originating element is too complex, as in the case of Phosphorous-30, the realignment time will prevent the investment of excess conversion energy into the magnetic current stream. The energy will be invested in the creation of a *mirror proto-neutrino* and the ejection of the new particle from the nucleus.

The excess energy from proton-to-neutron conversion has both been measured and calculated as "0.873 MeV."[83] The energy required to create the *mirror proto-neutrino* is "0.8602 MeV," as calculated by by Einstein's equation "$E=mc^2$." The remaindered energy between the known excess energy and *mirror P-neutrino* creation energy is the kinetic energy used to eject the particle from the nucleus. It is calculated as "12.82 keV." This translates to an ejection velocity of "0.1726" of the speed of light for a particle with the mass of the *mirror P-neutrino*. Joliot-Curies' cloud chamber recorded *mirror P-neutrinos* being ejected at 17.26% of the speed of light.

Despite the claims made by Joliot-Curies that they were recording "positrons," there was no evidence of the *"positron-electron annihilation"* which the model requires. Modern "positron" theorists claim that all the gamma emissions recorded by medical science's gamma cameras during Fluorine-18 beta decay are caused by *"positron*

---

81 *"The Quantum Geometric Model of the Neutron"* in *Four Dimensional Atomic Structure,* p.p. 71-93. Dawson, L. Op. Cit.
82 *A Quantum-Dimensional Model of Positive Beta Decay is revealed by Medical Science's 'PET Scan.'"* Dawson, L. The Snake River N-Radiation Lab. Op. Cit.
83 Ibid.

annihilations." Yet these *"annihilations"* are never seen when the positively charged particles emitted during proton-to-neutron conversions, are viewed with a cloud chamber. C.T.R. Wilson, the inventor of the cloud chamber, had bombarded his chamber with gamma radiation which produced a diffusion of tracks through the chamber. Nothing like Wilson's gamma-burst trails were seen during the Joliot-Curie experiment.

If the particles emitted during positive beta decay are not positrons, then what is the source of the gamma emissions recorded by medical science during the decay of Fluorine-18 to Oxygen-18? The answer is in the simplicity of the Fluorine-18 nucleus. The conversion of a proton to a neutron leaves three "in row" neutrons at the end of the magnetic-current circuitry . The geometry of the Fluorine-18 nucleus allows the transmission proton to simply flip the two excess neutrons into the corners of the triangular structure of the nuclear circuitry.

This flip is faster than the "pulse" defining the current[84] and, therefore, the decay reconfiguration of the nucleus does not interfere with the magnetic current flow. The excess energy from proton-to-neutron conversion is invested in the magnetic current and is output as gamma radiation by projection to the electrons.

The proton-to-neutron conversion within the simpler Fluorine-18 nucleus outputs excess energy as gamma radiation. The proton-to-neutron conversion within the more complex Phosphorus-30 nucleus outputs excess energy as lighter, positively-charged particles which do not annihilate to release gamma radiation. The "positron" model of positive beta-decay is thus eliminated, leaving the alternative Fermi *"electron-neutrino annihilation,"* as modified by Quantum-Dimensional mathematics, intact.

---

84 *"Calculation of Time Needed to Construct Mass and its Relationship to Energy"* in
   Four Dimensional Atomic Structure, p.80. Dawson, L. Op. Cit.

3. PARTICLE PHYSICS:
The Wittgensteinian Deconstruction of Nuclear Science

The arrival of "standard model" particle physics in the 1960s represented a deconstruction of scientific language and categorization. The objectively referenced atomic categories established by Ernst Rutherford in the first part of the twentieth century— the orbiting electron and the nuclear protons and neutrons— were objectively referenced as they were established by valid scientific research methods which identified categories of existent particles. They were replaced in the 1960s by artificial categories (hadrons/baryons/mesons etc.) without observable objective referents,categories which were imposed by social consensus. This transference from objectively referenced categorization to social consensus categorization represents the linguistic deconstruction introduced by the philosopher Ludwig Wittgenstein in the 1950s.

In the 1960s, the current author had witnessed the destruction of empirically-based social-science at Columbia University by the Wittgensteinian deconstruction of language:

"In 1967, I was given one of a limited number of faculty fellowships to the now defunct Bureau of Applied Social Research at Columbia University headed by the empiricist, Paul Lazarsfeld[85] who is known as the cofounder of mathematical sociology. Lazarsfeld was establishing a more scientific and less philosophically-based sociology at Columbia in conjunction with the theoretician Robert Merton, the founder of a new field called the 'sociology of science' for which Merton would later became the first social scientist to receive the National Medal of Science award in 1994.

"Columbia at the time was in turmoil, not only because of the Vietnam war, but academically as well. In conjunction with growing antiwar distractions, empirical science itself was under a philosophical assault which was posing itself as a new 'wave of the future.' The 'hottest' professor at the time was Peter McHugh, a former professional Hollywood actor who had entered academia and who was presenting lectures on the linguistic philosophies of Ludwig Wittgenstein. McHugh was attractive, well spoken and lectured with

---

85 *"Paul Lazarsfeld and Applied Social Research: Invention of the University Applied Social Research Institute"* Barton, Allen H. *Social Science History,* Vol. 3, No. 3/4 (1979), pp. 4-44

the motions and gestures of a trained actor. His were the most entertaining lectures in the department, and, needless to say, his lectures became a student favorite since they compared so favorably with the dryer presentations of less theatrically trained professors.

"McHugh was wrapping Wittgenstein in a pseudo social-science methodology called 'ethnomethodology.' However, he spent much less class time on this pseudo-methodology of sociologist Harold Garfinkel than he did on Wittgenstein's book 'Philosophical Investigations.' McHugh's 'witty' lectures drove home Wittgenstein's points from 'Philosophical Investigations'— a book written in 'witty' sophistries which McHugh's lecture style was imitating. McHugh taught Wittgenstein's idea that the nouns of our language did not and could not reflect objective reality. All nouns took their meaning by what McHugh called 'fiat' (by arbitrary decree). He was pushing the Wittgensteinian view that the nouns were subjectively imposed and were but socially agreed upon collections of unlike things without any connection to an objective reality.

"A common 'coffee house' debate among McHugh's students at the time were discussions comparing insanity with sanity. The strict Wittgensteinian view held that insanity and sanity could only be distinguished as private vs. collectively imposed meanings. Insanity was a system of privately imposed subjective meanings while sanity was a system of collectively imposed subjective meanings. Neither system could identify objective reality, so sanity was as far removed from truth as was insanity. It was concluded that the distinction was somewhat arbitrary.

"McHugh's 'show biz' lectures on Wittgenstein were destructive in a department with a strong commitment to the application of the scientific method. Empirical science is built upon the testing of factors governing cause and effect relationships. These factors must be 'externally existent' for these tests to have any validity. However, the philosophy of Wittgenstein, as popularized by McHugh's entertaining lectures, cast doubt among graduate students as to the validity of these testable factors as indicators of anything real. Increasingly, students began to view 'academic progress' as imposing revolutionary group-constructed redefinitions upon the obsolete and reactionary 'meanings' of past academic conventions.

"No one seemed to realize that science itself was under assault by this Wittgensteinianism; that without firm, objective meanings to our categories hypothesis testing simply could not occur. No one recognized or seemed to care that McHugh's Wittgensteinianism was undermining the scientific epistemology of the department.

"My fellow students were indifferent to the fact that they were laying an ax to the epistemological roots of western science itself. They were satisfied to dismiss empirical science with sniggering comments about a notorious failure on the part of Columbia's Lazarsfeld to correctly predict an outcome in one of his more famous studies, as if the failure of a hypothesis represented failure of the scientific method itself.

"The 'bible' on science at the time had became Thomas Kuhn's book, 'The Structure of the Scientific Revolution.' Kuhn reflected Wittgensteinian linguistic philosophy by proposing that scientific theories were merely functional belief systems which could be deserted for newer, more functional belief systems. None of these changing belief systems identified objective, underlying reality. The scientific ideation 'canopy' at any moment in history was always only a consensus belief ala Wittgenstein. Kuhn's argument was based upon the alleged replacement of Newtonian physics by relativity and quantum mechanics in the early twentieth century.

"Bertrand Russell has said that Wittgenstein intimidated him when Russell had him as a student and Wittgenstein had come to be considered the most influential philosopher of the twentieth century by many. Because of this massive reputation, no alternative was ever given to Wittgenstein's view that human categories are composed of unlike elements; that our categories are groupings of things which are placed together simply by McHugh's 'fiat.' If this is true, if our categories do not identify a factual commonality but are only an arbitrary grouping by human decree, by 'consensus,' then using those categories for scientific description is not possible.

"This viewpoint, taught by a socially attractive professor and accepted uncritically by my student peers, produced in me an existential crisis. What was the point of education if knowledge did not identify truth about things, but only a temporary belief about things; a belief which could change with time and fashion? I remember standing in the Columbia library stacks paralyzed and unable to pursue topical research because I had been denied by Wittgenstein the critical tools needed to evaluate the articles I might read. I remember watching, with envy, a friend pursue library research for his orals, with no compunction and with vigor, simply because he accepted that understanding something which is collectively believed is the same as acquiring knowledge. For him, academic science had become a social process with which he felt comfortable. He felt absolutely no requirement for a sound epistemological underpinning to knowledge. "[86]

---

86 *"Why I don't have a Ph.D and, by extension— 'what does your Ph.D actually*

During the same time frame in which empirically-founded social-science was being deconstructed at Columbia University, a similar process was occurring in nuclear science at Stanford University. In 1968, Richard Feynman was using the Stanford Linear Accelerator[87] to deceptively establish a linguistically deconstructed model of neutrons and protons.

Feynman alleged he was "bouncing" accelerated electrons off nuclear protons to establish that the protons were composed of smaller particles. Feynman was making this claim in support of an artificial re-categoriz-ation of the nuclear particles in conformity to a new social-consensus model (the quark-based "standard model"). Feynman was making this claim to establish the re-categorization despite the fact that it had been known for forty years that accelerated electrons do not "bounce" off nuclear protons. It was known that they cannot penetrate the electron orbital field and are scattered at x-ray orbital distances[88].

The "standard model" of atomic particles is a Wittgensteinian linguistic deconstruction which occurred during the 1960s. It removed particle categories as *objective referents* to atomic functionality, replacing them as artificial categories the belief in which could benefit the careers of participants in the scientific collective. Ludwig Wittgenstein taught that the categories of objects subsumed by our words do not share a functional reality but are *"objectified"* only by social agreement.[89] Things are in the same category only because we agree they are the same, not because they have any objective "sameness." This removes word meanings from having any objective truth and makes them "real" or "true" only by a vote.

The protons and neutrons were no longer the nuclear cousins they had been proved to be; that the neutron was not made by the merging of an electron with a proton, with an electron mass gain of "2.5" *times*; that the neutron ejected from the nucleus decays back to a proton in an average "6

---

mean?'" Dawson, L. The Snake River N Radiation Lab.
http://www.paradigmphysics.com/phdata.pdf

87 *"Modern View (Standard Model) timeline: 1964 - present"*
http://www.particleadventure.org/other/history/smt.html
88 See Chapter 2.
89 *The Death of Reality,* p.p. 39-72, *"A 'Stealth Marx' of Political Unreality."* Dawson, L. The Paradigm Company, 2015. ISBN. 978-0941995368

1/3 minutes" by the ejection of an electron.[90] By social agreement it was now known that both protons and neutrons were composed of three "quarks" and the particles comprised an artificial particle category called "hadron/baryons."

The existence of these "quarks" had never been demonstrated until Richard Feynman said he had "bounced" electrons off protons using the Stanford Linear Accelerator. Feynman claimed that the scatter data from "bouncing" the electrons off the nuclear protons enabled a measure of the smaller "quarks" which the "standard model" now asserted the proton was composed. Unfortunately, the use of the Feynman "bounce" method revealed quark masses which were 153 *times* too small to explain the proton's mass[91].

The Feynman deceptive technique, which identified impossibly small "quark" masses, did not disturb anyone. The "quark" was a Wittgensteinian linguistic transformation which disconnected the artificial category from any objective referent and invested it as a subjectively imposed collective meaning. The "quark" was "'bound" within the proton and existed by a social agreement with little need to supply an objective referent. George State University's HyperPhysics page on "quarks" reveals this indifference to the objective measurements of "quarks:"

> *"The* [inadequate quark] *masses should not be taken too seriously, because the confinement of quarks implies that we cannot isolate them to measure their masses in a direct way. The masses must be implied indirectly from scattering experiments."*[92]

Inadequate measurements of "quark" masses using the deceptive Feynman accelerator technique— the only direct proof or *"objective referent"* for "quarks"— should not be taken *"too seriously"* because the "quarks" are confined and cannot be measured in *"a direct way."*

In the face of contradictory data, "quarks" existence could still be imputed by the collective assertion of their reality. "Quarks" required no

90 *"Neutron decay rates decrease under the influence of a magnetic field proving the neutron's 'magnetic current' function,"* p.4. Dawson, L. Op. Cit.
91 Ibid.
92 *Quarks.* Georgia State University HyperPhysics, Dept. of Physics and Astronomy. http://hyperphysics.phy-astr.gsu.edu/hbase/particles/quark.html

objective referent to confirm. The "quark standard model" of nuclear particles, as established in the 1960s, represented a Wittgensteinian deconstruction of scientific language which removed any requirement for categories to possess an "objective referent." .

The removal of any "objective referent" requirement for the "quark" categorization system is proved by the second category, the "hadron/mesons." "Mesons" were said to be composed of a "quark" and an "anti-quark" of a different "flavor." This alleged particle was composed of an unprovable "quark" and antimatter. It required no observation to establish as "fact," only the belief in its existence by the scientific collective.

The only "objective observation" which might help establish the "anti-quark" component of the "meson" was the measurement of the "positron" or the "anti-electron" in the 1930s.[93] This "antimatter " particle, which supposedly annihilated an electron in a high energy burst upon contact, were apparently recorded by Frédéric and Irène Joliot-Curie during a study of the conversion of protons to neutrons in 1934.

Although the particle emitted during the positive beta decay of Phosphorus-30 was positively charged (the electron is negatively charged) and close to the mass of an electron (a small fraction of proton and neutron masses), the particle did not display the a*nnihilation reaction* predicted for "antimatter." The particle was actually undergoing the non-energetic *"electron-neutrino annihilation"* foreseen by Enrico Fermi. The particle recorded by the Joliot-Curies was actually the *mirror P-neutrino* identified by the quantum-dimensional model of proton-to-neutron conversion.[94]

The very existence of an "antimatter" particle is questionable since it is founded solely upon the alleged Joliot-Curries recording of the anti-electron "positron"— a particle which, it is alleged, is emitted during positive beta decay, but which has never been shown to annihilate as predicted. Nonetheless, a second category incorporating "quark anti-matter" was built "exo-data" (outside data) and accepted as "true."

The deceptive use of particle accelerators, as introduced by Richard Feynman, then became the standard "methodology" by which the new

---

93 See Chapt. 2, .p.p. 30-44.
94 Ibid.

particle categorization system might be "justified." Accelerator observations became Wittgensteinian and disconnected from "objective referents." This followed earlier use of deceptive cosmic ray observations.

The deceptive Wittgensteinian underpinning all of particle physics can be seen in the development of the "pion" or "pi-meson" concept. Various "pions" are supposed to be composed of a "quark" and a different-flavored "anti-quark." These are said to be responsible for proton/neutron conversions. It is a claim made in direct contradiction of the research into the relationship and natures of the proton and neutron; the research conducted by Cambridge's Rutherford Group which established the field of nuclear physics.

The neutron was shown by the Rutherford Group to be composed by a proton merged with an electron, the merger of which increased electron mass by approximately "2.5 times."[95] Beta decay is the process of conversion; negative beta decay being the ejection of an electron from a neutron (called "beta radiation") to create a proton and positive beta decay being the merging of an electron with a proton to create a neutron which is accompanied by an energy gain of "0.873 MeV." [96]

The alternative to the neutron model as composed of electrons merged with protons (the model proved by the Rutherford Group) was the conversion of protons to neutrons by the speculative *"interactive particle,"* the *"hadron/meson/pion."*

By the time the proposed "pion" had been incorporated into the 1960s "quark" classification system as a "hadron/meson" it had undergone multiple generations of exo-data theoretical revision (Yukawa, 1936, Powell et. al., 1947, Lattes and Gardner, 1948, Goldstone, 1961)[97].

These "quark"/"anti-quark" pairs were said to substitute for the proved neutron/proton interactions from the merging and ejection of electrons as the primary source of neutron/proton conversions.

---

95 See Chapt. 2, p.p. 25-30
96 *A Quantum-Dimensional Model of Positive Beta Decay is revealed by Medical Science's 'PET Scan.'"* Dawson, L. The Snake River N-Radiation Lab. Op. Cit.
97 *"Pions"* Wikipedia. https://en.wikipedia.org/wiki/Pion#cite_note-3

"Interpretations" of cosmic radiation imprinted upon photographic plates were used to "prove" the existence of the particle. It was assumed that high energy cosmic radiation (mostly high-velocity nuclear material) interact with the atmosphere to produce the particles.

> *"Photographic emulsions based on the gelatin-silver process were placed for long periods of time in sites located at high altitude mountains, first at Pic du Midi de Bigorre in the Pyrenees, and later at Chacaltaya in the Andes Mountains, where the plates were struck by cosmic rays[98].*

> *"After the development of the photographic plates, microscopic inspection of the emulsions revealed the tracks of charged subatomic particles. Pions were first identified by their unusual 'double meson' tracks, which were left by their decay into a putative meson."[99]*

.

Actually, it was tracks of "doublet electrons" which the plates revealed. Photographic plate emulsions are sensitive to high energy electrons which is the principle by which electron microscopes operate. The "track" of an electron upon a photographic plate is much more refined than that of a lightwave which is the reason electron microscopes can resolve detail at much greater magnitudes. Detecting electron cosmic ray "tracks" would require a *"microscopic inspection"* which, in this case, revealed *"unusual 'double meson* (sic)' *tracks."* The researchers took the rare cosmic radiation *"'double...' tracks"* to *"mean"* they revealed *"pion/mesons"* rather than an electron doublet.

The "double tracks" on the plates exposed to cosmic radiation are consistent with the "light doublets" which appear on photographic plates of light split by a prism (spectrographs). Both energized hydrogen and sodium display "double lines." It has been proved, mathematically, that the double lines are output by electron orbits which are slightly offset.[100] These "nested doublet" electrons are closely bound together and could be emitted together as cosmic radiation doublets. They would produce the "double tracks" recorded by Powell *et.al.* In 1947.

---

98 Cecil Powell, César Lattes, Giuseppe Occhialini (University of Bristol)
99 *"Pions"* Wikipedia.   https://en.wikipedia.org/wiki/Pion#cite_note-3
100 *Four Dimensional Atomic Structure,* p.p. 41-48. Dawson, L. Op. Cit.

An actual 1947 "double track" on a photographic plate.[101]

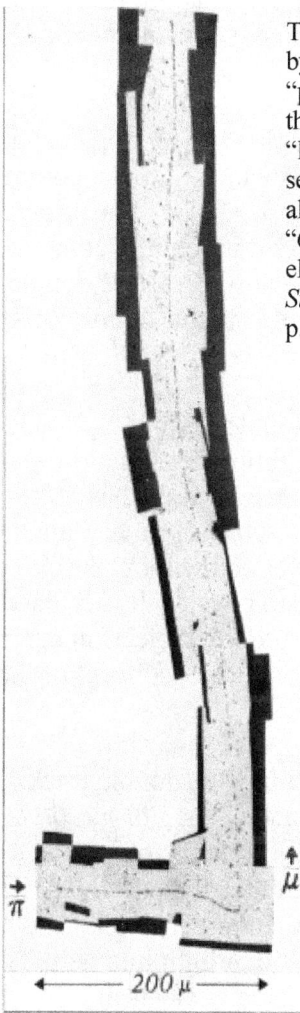

"Pion" Interpretation of Data not Possible

The actual cosmic radiation "double track" recorded by Cecil Powell *et. al*. In 1947 could not be the "pion decay" claimed for it. Notice the direction of the "pion" (designated as "π"). It is said to decay to "light" at an angle of dispersion which sends the second "light" track counter to the direction of the alleged "pion." The angle of dispersion measures as "60°" which is the standard quantum angle of the electron orbitals. (SEE: *Four Dimensional Atomic Structure; "Four Dimensional Orbital Structure,"* p. 168. Op. Cit.)

Angle of dispersion is 60°± 1°.

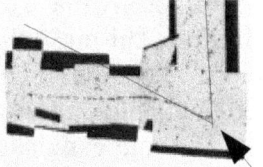

Direction of impact
From electron doublets

Force-Vector and Dispersion Angle prove Doublet

The angle of dispersion for the two tracks recorded by Powell prove that they are caused by electron doublets striking a photographic emulsion atom and being scattered at the quantum orbital angle. Both the angle of scatter and the obvious vector of force prove the tracks are by electron doublets scattered on impact with the emulsion.

← 200 μ →

Primary meson track, π, with secondary light particle, μ

The concept of *"antimatter,"* which is said to partially compose Cecil Powell's alleged *"pi-meson"* or *"pion,"* was first put forward by Paul

101 *"CECIL POWELL: Fragments of Autobiography,"* p.26. University of Bristol, 1987
http://www.bristol.ac.uk/physics/media/histories/12-powell.pdf

Dirac to explain spectrographic light doublets known as the "Zeeman Effect."[102] The alternative to Dirac's "antimatter" explanation for Zeeman light doublets is the slightly offset orbitals revealed by quantum-dimensional mathematics.

The doublets which appear on the spectrographs[103] of hydrogen and sodium are explained by two electrons in orbitals which output a common "root" wavelength from which they are slightly offset. The "fine structure" of the root wavelength on the spectrograph shows it is actually composed of two line; one slightly above and one slightly below[104]. These doublet spectrographic lines are caused by doublet electrons in the slightly offset orbits.

It is a curious scientific irony that Powell's double cosmic-ray tracks—which he alleged proved the "pi-meson" and its antimatter function—actually proves the electron doublet theory; the theory which has completely eliminated the Dirac *"antimatter equation"* as an explanation for the Zeeman Effect. Without explanative power and without empirical confirmation, Dirac's *"antimatter"* has no place in physics. It is replaced by quantum-dimensional mathematics and their proposed electron orbital doublets. As falls the concept of "antimatter" so falls the concept of the "pion" which is based upon it.

Powell *et. al.* imposed an artificial meaning upon the double tracks which should have been eliminated by the data itself. The *"pi-meson (π)"* is said to decay to a *"light particle (μ)"* which is ejected at 60° to the track of the *"pi-meson or pion."* However, the track of the alleged *"light particle"* is bent away from its vector of travel then re-bent back to its original vector. This is completely uncharacteristic of light interacting with matter, but can be characteristic of a high velocity electron interacting within an emulsion.

Electrons with counter-clockwise spin would place a "left-leaning" force against the declining force supplied by the electron's velocity

---

102 Chapter 2, p.p. 33-36.

103 Photos of light-wavelengths spreads under the influence of a prism.

104 *"Pieter Zimmerman's use of a magnetic field to modify the naturally occurring light doublets for sodium has revealed the means by which electron orbital subshells acquire multiple electrons"* in Four Dimensional Atomic Structure, p.p.41-48. Dawson, L. Op. Cit.

through the emulsion. When the force of velocity weakens enough, the force of counter-clockwise spin can "steer" the electron to the left.

Both legs of the Powell double tracks demonstrate this inclination towards a "left steerage." Interestingly, the left leg, has an initially darker track (from the vertex of dispersion) indicating temporary movement through denser emulsion. The left leg "steerage" occurs at a shorter distance from the angle vertex than does the right leg with a lighter track indicating movement through less dense medium.

Cosmic radiation which is tangential to the earth's surface (so-called "dawn" radiation), radiation which is actually composed of an electron doublet, intersects the surface of the Powell photographic plate to create the anomalous double tracks. These anomalous tracks have all the characteristics of electron scatter upon impact with an emulsion atom. Nonetheless, the researchers ignore the fact that the tracks have the signature of electron scatter.

Instead they interpret the tracks to "mean" a speculative particle which contains a component of speculative "antimatter." They force an interpretation to give credence to a theoretical interactive particle which substituted for the research-founded explanation for proton-to-neutron conversion. The speculative *"pion"* and its *"anti-quark"* replace electron merger and ejection with protons and neutrons as the cause of neutron-proton conversions. The interpretation thus imposed upon the double track data is a Wittgensteinian linguistic transformation of that data.

By the 1960s, particle accelerators had replaced cosmic radiation photography as the preferred form of "data support" for the emerging particle theories. But the deceptive technique invented for cosmic radiation photography was also applied to accelerator data. Specifically, anomalous (not fully understood) data is ascribed to a theoretical particle. A Wittgensteinian linguistic transformation of the data is imposed.

The most recent example of anomalous accelerator data being made to "fit" consensus particle theory is the case of the alleged *"Higgs Boson"* or so-called *"God particle."* A *"boson"* is a pseudo, mathematically-proposed particle which is said to contain "force" or to provide some other explanation for physical interactions which are not understood. The name was proposed by Paul Dirac, the man who proposed "antimatter" to explain the Zeeman Effect.[105]

In 2012 the CERN particle accelerator in Geneva Switzerland recorded an anomaly during ultra high-energy particle collisions which they said fit the theoretical *"Higgs Boson"* particle.

> *"The CMS [CERN] experiment team claimed they had seen a 'bump' in their data corresponding to a particle weighing in at 125.3 giga electron volts (GeV) - about 133 times heavier than the protons that lie at the heart of every atom."*[106]

The CERN collider accelerator is the most powerful in the world, with electromagnetic acceleration energy which are well beyond those required to accelerate a proton to the speed of light.

> *"The highest particle energies ever achieved have been produced with the Large Hadron Collider (LHC)—a superconducting proton synchrotron at CERN in Geneva—which accelerated protons to 1.18 tera-electron volts (TeV; one trillion electron volts)."*[107]

Energies applied to protons of "1.18 TeV" are "2515" *times* the energy required to accelerate the proton to the speed of light. Colliding protons accelerated to near-light speeds will have twenty-five hundred *times* the force behind them as that provided by their velocities alone. This force may be enough to collapse the gravitational system, as well as annihilate the particle. CERN's *"125.3 GeV bump"* may have recorded the collapse.

CERN's *"1.18 TeV"* accelerator is providing similar energies to the recorded gamma radiation bursts which accompany the ejection of matter from large black holes at the centers of galaxies. These black holes can emit tera-electron voltage (TeV) gamma radiation bursts and are known as

---

105 Notes on Dirac's lecture *Developments in Atomic Theory* at Le Palais de la Découverte, 6 December 1945, UKNATARCHI Dirac Papers BW83/2/257889. See note 64 to p. 331 in "The Strangest Man" by Graham Farmelo

106 *"Higgs boson-like particle discovery claimed at LHC"* Paul Rincon, Science editor, BBC News website, Geneva: http://www.bbc.com/news/world-18702455

107 "Synchrotron Physics," Editors of Encyclopædia Britannica. http://www.britannica.com/technology/synchrotron#ref866740

"blazers."[108]

Quantum-dimensional mathematics have identified what black holes are and why they can emit *"blazer"* gamma radiation. Black holes have been shown to be the gravitational focal points of large interstellar formations such as galaxies.[109] They are gravitational focal points which have flattened quantum space and collapsed that space to the radius of the disk.[110] This creates a *"singularity-event horizon"*[111] at the periphery of the disk.

Matter pulled across the *"singularity-event horizon"* onto the surface of the disk loses all definition of mass and is instantly accelerated at 90° away from the face of the disk[112] to velocities which can be greater-than-light speeds. The ejection energy is a function of the diameter of the rift in space.

This creates immense potential energy. The matter reinters quantum space to reacquire gravity and the spacial speed-of-light restriction. The matter may reenter gravity-influenced space with greater potential energy than can be expressed by the conversion of its mass to energy. The energy has been pushed beyond conversion energy by the particle's interaction with the black hole. The boosted particle can emit gamma radiation greater than "1065 *times"* (>1 TeV for a proton) its annihilation energy.

The CERN accelerator reversed this process. Protons could be accelerated under "1.18 TeV" of electromagnetic energy; energies similar

---

108 *"Ultra High Energy (UHE) gamma emissions are associated with what have become known as 'blazars.' They are 'jets' associated with the center (nuclei) of galaxies which emit UHE gamma which intercept the earth when the 'jets' are in a direct line of sight with our view."* in *Four Dimensional Atomic Structure,* p. 158. Dawson, L. Op, cit.

109 *"A Black Hole as an Open Energy System,"* Dawson, L. SRNRL Astrophysics Module 501.   http://paradigmphysics.com/black-hole-book.pdf

110 VIDEO: *"Black Hole Open Energy Lecture using the CHANDRA J-1550-564 Black-Hole Jet X-Ray Data" Dawson, L.*
     https://www.youtube.com/watch?v=PkUM74D0Gus

111 *"Creation-field cosmology: A possible solution to singularity, horizon, and flatness problems"* Physical review D: Particles and fields 32(8):1928- 1934 October 1985

112 VIDEO: *"Black Hole Open Energy Lecture using the CHANDRA J-1550-564 Black-Hole Jet X-Ray Data" Op. Cit.*

to black hole *"blazer"* emission energies. When forced into collision at near-light speeds, the protons were being pushed by over 2000 times the energy contained in their velocity momentum. The "pushed momentum" had collapsed quantum space during the collision and momentarily opened a black-hole-like *"singularity-event horizon."* The particle could acquire excess potential energy within a temporary "hole" which had been caused by the forced collapse of quantum space.

When the temporary "hole in space" broke down, the CERN researchers recorded an energy burst which was "133 *times*" the conversion energy of the proton. This they interpreted to "mean" the existence of the alleged *"Higgs-boson"* particle of pure theoretical (pseudo-mathematical) invention.

There is no logical mechanism to explain how a proton pushed by an electro-magnetic field to two thousand times the force of its maximum velocity can be collided to produce a particle "133 *times*" its mass. This is the problem which "standard model" particle physics avoids with its *"Higgs Boson"* speculation. In the collision, the proton has somehow converted to the *"Higgs Boson"* which, it is said, has a mass of "133 *times*" the mass of the proton.[113]

The problem of the mass difference between the input proton and the allegedly outputted *"Higgs Boson"* is simply eliminated as a problem by a Wittgensteinian linguistic transformation. A Wittgensteinian language deconstruction removes categories from any objective referent and supplies "meaning" to the categories as a socially-enforced belief system. Thus, the "objective referent," in this case being the the problem of proton-to-Higgs-Boson mass conversions, is removed as a problem for logic to resolve. The "mass problem" can be simply ignored altogether because another meaning for the event is established as a "fact" by the socially enforced "Higgs Boson" interpretation of the event. By removing the event category as an objective referent, its "interpretation" is left completely to consensus supplied "meanings" with the loss of all logical capacity.

However, the *"bump"* which the CERN researchers had recorded of *"125.3 GeV"* was actually a *"blazer-like burst"* created when accelerator energies had reached black-hole "blazer" levels. This actual explanation of

---

113 *"Higgs boson"*   Wikipedia.   https://en.wikipedia.org/wiki/Higgs_boson

the CERN data retains the event as an objective referent.

The excessive electromagnetic field "push" of the proton at the collision point collapses quantum space in front of the proton and opens a temporary *"singularity-event horizon"* which allows an increase in kinetic potential energy greater than the annihilation energy of the proton.

A *"singularity-event horizon"* is a characteristic of the edge of a black-hole disk across which the potential time energy which establishes all space no longer exists. The potential energy component of time is the force by which all space and matter have been created and is the force which sustains quantum space.[114] The *"singularity"* is the creation of space by potential time energy (time-enforced space). The black-hole *"horizon"* is the peripheral boundary across which time-enforced space exists but behind which it does not. Behind the boundary matter cannot exist and will be ejected if it trespasses.

The *"radius of rift"* which the black-hole's gravitational focal point imposes upon quantum space will determine the force of ejection imposed upon matter which crosses the *"singularity horizon."* In the case of small black holes, such as CHANDRA'S J-1550-564 black hole, matter is emitted with sub-annihilation energy gains.[115] For great black hole rifts, like those at the center of galaxies, matter crossing the "singularity horizon" are ejected at many times the annihilation energy of matter.[116]

For the CERN particle accelerator, the data from which has been misinterpreted as a *"Higgs Boson,"* the number of times the proton's annihilation energy is multiplied is a measure of the *"radius of rift"* in time-enforced space[117]. This *"rift"* has been produced by the excess

---

114   VIDEO: *"The Quantum Dimension......."* on YouTube by The Snake River N
        Radiation Lab   https://www.youtube.com/watch?v=ArLRKvwy4_8
115   Velocities less than the speed of light. SEE VIDEO: *"Black Hole Open
        Energy Lecture using the CHANDRA J-1550-564 Black-Hole Jet X-Ray Data"*
        Dawson, L   .https://www.youtube.com/watch?v=PkUM74D0Gus
116   *"Ultra High Energy (UHE) gamma emissions are associated with what have become
        known as 'blazars.' They are 'jets' associated with the center (nuclei) of galaxies
        which emit UHE gamma which intercept the earth when the 'jets' are in a direct
        line of sight with our view."* in *Four Dimensional Atomic Structure,* p. 158.
        Dawson, L. Op, cit.
117   *"The Theory of Time-Enforced, Four-Dimensional Space"* in *The Quantum
        Dimension,* p. 112. Dawson, L. The Paradigm Company, 2015.
        ISBN: 978-1517233099

electromagnetic force which the accelerator has imposed upon proton momentum. The accelerator electron voltage is at *"blazer"* levels associated with large black holes at the center of galaxies (1.18 TeV).

To reiterate, the CERN researchers produced an anomalous observation during the collision of protons under accelerator energies which approached those of black-hole *"blazer"* levels. A *"bump"* in the recorded energy from one of the proton collisions revealed an annihilation energy (125.3 GeV) 133 *times* the annihilation energy available to a proton (0.9838 GeV). This record was interpreted to "mean" the annihilation of the speculative *"Higgs Boson"* particle.

This interpretation was possible by removing the event from any objective referent as to its cause and effect. A Wittgensteinian transformation made the event "mean" a *"Higgs Boson"* annihilation without reference to how a proton collision under tera-electron-voltage acceleration could result in an alleged *"Higgs Boson"* annihilation.

The alternative explanation, as a *"rift in quantum space"* which boosts proton annihilation energy, reinstates cause-and-effect. It does so by identifying the accelerator energy which was applied as consistent with the black-hole *"blazer"* energies which are known to be caused by such rifts in time-enforced space. The multiplication of proton annihilation energy by the CERN accelerator is identified as consistent with that resulting from the expulsion of matter from high-energy black holes.

THE DESTRUCTIVE HISTORY OF PARTICLE PHYSICS
Ignorant Interpretation and Deliberate Deception

Artificial particle physics originated in the inability of primitive quantum mechanics to explain the anomalous "Zeeman Effect" — the multiplication, under a magnetic field, of the light doublets known as the "D lines" which appear on the spectrograph of sodium. In 1931 Cambridge's Paul Dirac proposed that atomic particles had a negative energy or "anti-matter" state as an explanation for the Zeeman Effect.

The Dirac equation which proposed an *"anti-electron"* had no basis in empirical observation. However, in 1933 the American astrophysicist Carl Anderson observed a cosmic radiation particle in the Mount Wilson

Observatory's cloud chamber which had a positive charge but a fraction of the mass of the positively charged proton; a mass closer to that of the electron. This particle was designated as Dirac's *"anti-electron"* even though the cloud chamber gave no evidence of the annihilation such anti-matter was supposed to undergo.

Without the reformation of primitive quantum mechanics by quantum-dimensional mathematics— which would occur 80 years later— Anderson could not recognize that his supposed "anti-electron" was actual a *"mirror P-neutrino"* with a 2/3 positive charge and mass of 2.5 *times* the mass of the electron. Quantum-dimensional mathematics has revealed that such a particle can be output during the positive beta decay of a proton to a neutron.

In 1933 Frédéric and Irène Joliot-Curie discovered that Anderson's alleged "anti-electron" appeared in a cloud chamber during the positive beta decay of Silicon-30 to Aluminum-30. However, the observed particles fit the *"mirror"* particle emission from positive beta-decay which is predicted by quantum-dimensional mathematics. The observation did not fit the "anti-electron" proposed by Dirac.

Dirac's *"antimatter"* was proposed to explain atomic process. The energy gained through *"electron/anti-electron annihilation"* was the reason Dirac proposed the existence of the anti-particle. In the electron rich environment of Joliot-Curies' ionized water-vapor cloud trails such annihilations should have been commonplace. However, no tell-tale high energy gamma radiation bursts were recorded by the cloud chamber. Instead, the particles seemed to disappear at the end of the cloud trail, possibly being converted into a massless neutrino by *"mirror-P-neutrino/electron annihilation."*

On contact with an electron, the *"mirror Proto-neutrino"* converts into a neutrino without the annihilation energy-burst allegedly associated with Dirac's *"anti-electron."* The *"anti-electron"* annihilates in a gamma radiation burst, while the *"mirror P-neutrino"* annihilates quietly[118].

Despite the lack of *antimatter* annihilation data in either the Anderson

---

118 SEE: *"Fermi's Theory of Beta Decay, A complete English translation."* (Op. Cit.) for Fermi's description of the difference between *"electron/neutrino"* annihilation and *"electron/anti-electron"* annihilation.

or the Joliot-Curie cloud-chamber records, the positively charge particles with near electron mass were identified as *"anti-electrons"* or *"positrons."* Even in the face of nonsupport from the Joliot-Curie experiment, gamma radiation during positive beta decay would go on to be credited to *"electron/anti-electron annihilation"* from *"positron"* emissions.[119]

The alleged "positron" was the first non-constituent particle proposed for the atom and was used to establish the existence of Dirac's *"anti-particles;"* particles which annihilated on contact with their positive version. Dirac's *"antimatter mathematics"* were credited with successfully predicting the "anti-electron" despite the fact that both cloud chamber observations of these positively charge particles of near-electron masses did not actually support the theory. The "positron" interpretation was accepted by consensus. No alternative *"non-annihilating particle"* was sought. The *"mirror P-neutrino,"* which can be emitted during positive beta decay, could not be discovered because it was excluded by the exo-data acceptance of the *"positron."*

Dirac's *"anti-particle"* became a *"fact."* The forced interpretation of data to credit speculative particle theories became endemic. These forced interpretations prevented actual discoveries which could benefit our knowledge of atomic process.

In 1947, Powell *et. al.* forced interpretation of double tracks from cosmic radiation events recorded on photographic plates. He claimed the tracks proved the existence of the *"pi-meson"* or *"pion,"* a theoretical particle containing an *"anti-particle"* component which, it was alleged, facilitated beta decay.

An authentic reading of the track record, however, proves that the Powell interpretation could not possibly be true. The tracks bear the clear signature of particle scattering, not *"pi-meson annihilation into a light particle."* The forced interpretation prevented discovery of the tracks as scattered electron doublets; a discovery which would have facilitated authentic knowledge of atomic orbital structure and the method by which

---

119 SEE: *"A Quantum-Dimensional Model of Positive Beta Decay is revealed by Medical Science's 'PET Scan' ,"* Dawson, L. SRNRL.
http://paradigmphysics.com/PET-scan-pos-beta-decay-theory2.pdf

orbital structure manages orbital electron capacities.[120]

In the 1960s, A Wittgensteinian linguistic transformation was applied to nuclear particles. In his 1953 book, *Philosophical Investigations,* Cambridge's Ludwig Wittgenstein had taught a method of deconstructing word categories by detaching them from their objective referent. A group of objectively unlike objects could be put together and made to "mean" the same thing by a social-consensus use of the word category. Things can "become" the same thing by pure social consensus.

This Wittgenteinian transformation was applied to the known nuclear particles. The observed nuclear particles— the neutrons and protons— were lumped together with the purely speculative particles such as the *"pi-meson"* into a new consensus-imposed category system. In this way the speculative particles were made to seem that they shared "reality" with known particles.

The method by which the objectively known particles could be made to share the same category with the speculative particles was the invention of the *"quark."* The *"quark"* was completely a linguistic invention developed solely so the observed protons/neutrons could share the same word category with *"pi-mesons"* and other speculative particles.

Neutrons, protons and *"pi-mesons"* were all said to be *"hadrons"* because they were all composed of *"quarks"* and/or *"anti-quarks."* The objectively known particles were said to be composed of three *"quarks"* and were said to be *"hadron/baryons."* The speculative particles were said to be composed of one *"quark"* and one *"anti-quark"* and were designated as *"hadron/mesons."* The new linguistic designations were socially enforced, so much so that the CERN high-energy proton collider is currently known as the *"Hadron Collider."*

The initial obstacle confronting the Wittgensteinian transformation of objectively known particles into a category shared with the speculative particles was the credibility of the *"quark."* This was resolved by Richard

---

120 *"Pieter Zeeman's use of a magnetic field ....revealed the means by which electron orbital subshells acquire multiple electrons."* in *Four Dimensional Atomic Structure,* p.p. 41-48. Dawson, L. Op. Cit.

Feynman's deceptive use of the Stanford University Linear Accelerator.[121]

> "[Using] *the Stanford Linear Accelerator, in an experiment in which electrons are scattered off protons, the electrons appear to be bouncing off small hard cores inside the proton. James Bjorken and Richard Feynman analyze this data in terms of a model of constituent particles inside the proton...*"[122]

Feynman's announcement was pure deception because it had been known for 50 years that fast-moving electron's don't *"scatter"* off nuclear protons but are scattered at orbital distances associated with x-ray.[123] Feynman *et al* simply pretended they were *"bouncing"* electrons off protons. They did so in order to say that the lower scatter energies associated with x-ray orbital distances *"proved"* that the electron was not *"bouncing"* off the whole mass of the proton, but only a *"small hard core inside the proton."*

Unfortunately for the Feynman deception, the lower orbital scatter energies, as measured by linear accelerators, translated to a *"quark"* mass value which was much too low. In order to make the *"quark"* linguistic invention viable as means to incorporate proton/neutrons into the *'hadron"* classification system, *"quark"* masses must be approximately "1/3" of proton/neutron masses. However, accelerator electrons scattered at x-ray orbital distances were only translating into *"quark"* masses which were a fraction of what they needed to be. Electron scatter data translated to alleged *"quark"* masses which were less than 1% of what was required (0.659% - 0.4% of required mass).[124]

Ultimately, it did not matter that the Feynman deception could not identify actual *"quark"* masses.[125] The *"quark"* and its classification system, which integrates known particles with speculative ones, is a Wittgensteinian transformation. It deconstructs word categories by removing them from having an objective referent to applying meaning by

---

121 See Chapt. 2, p.p. 20-22.
122 *"Modern View (Standard Model) timeline: 1964 - present"*
http://www.particleadventure.org/other/history/smt.html
123 See Chapt. 2, page 20 and forward.
124 Chapt. 2, p. 24.
125 Chapt. 2. p. 20

social consensus and social consensus alone. Science was undisturbed by the fact that Feynman's deceptive electron scatterings did not reveal authentic quark masses because *"the masses should not be taken too seriously, because the confinement of quarks implies that we cannot isolate them to measure their masses in a direct way"*[126]

"Quarks" are *"confined"* within the proton/neutron masses and cannot be measured *"in a direct way."* *"Quarks,"* therefore, have no objective referent. The proof of their existence by the Feynman deception is accepted by social consensus even though it is admitted that Feynman electron scattering cannot actually measure the alleged *"quarks."* There is no inconsistency in accepting Feynman electron scattering as "proof," then acknowledging that Feynman electron scattering cannot actually measure the *"quark."* The *"quark"* is a language deconstruction with no need for an objective referent, only a collective belief. The 1968 Feynman "proof" is actually irrelevant. *"Quarks"* are made real during conversations, by their acceptance during conversations between adherents.

Why should science accept the *"quark"* linguistic invention as a substitute for empirically founded atomic research? The answer, I am afraid, is corruption. Academic careers can be advanced more by the acceptance of speculation than by the rare cases of actual discovery. But that may not matter to you because you may also accept personal corruption as a means to selfish interests. You may also be an enemy of truth.

The *"quark"* linguistic invention allowed academics to package speculative particle theories in the same categorical system with the particles known to compose atoms. The known particles were transposed from having "atomic function" to having "membership" in an artificial category system which also included the speculative particles which could further academic advances. Anomalous observations were made to "mean" evidence for speculative theory. The proton annihilation in CERN's *"Large Hadron Collider,"* with an energy gain of *"133 times"* the conversion energy of the proton[127], was made to *"mean"* evidence for

---

126 *"Quarks"* at Georgia State University Hyperphysics.
   http://hyperphysics.phy-astr.gsu.edu/hbase/particles/quark.htm
127 p.p. 58.

the theoretical *"Higgs[128] boson."*

Ignored were the real-world data from black-holes which indicated the proton-annihilation energy increase could be due to the production of a *"singularity event-horizon"* and the collapsing of quantum space. Instead of identifying similar energy events from *"blazer"* black holes, CERN researchers applied completely speculative theoretical physics to the event.

> *"In the 1960s, [Higgs] proposed that broken symmetry in electroweak theory could explain the origin of mass of elementary particles in general and of the W and Z bosons in particular. This so-called Higgs mechanism, which was proposed by several physicists besides Higgs at about the same time, predicts the existence of a new particle, the Higgs boson, the detection of which became one of the great goals of physics."[129]*

Without any objective proof, it was speculated that *"electroweak theory could explain the origin of mass of elementary particles* (neutrons-protons) " Based on the exo-data *"electroweak theory"* a new particle, the *"Higgs boson"* was postulated as an explanation of the origins of particle mass. Again actual particles, the neutrons and protons, were brought into categorization with purely speculative particles, removing them from objective reference and placing them in a category established by collective belief. Again, it was a Wittgensteinian deconstruction of language. A Wittgensteinian interpretation of the CERN data was accepted by consensus, unmodified by its obvious connection to the *objective* gamma energy measurements from black-hole *"blazers.[130]"*

---

128 Peter Ware Higgs (born 29 May 1929) is a British theoretical
     physicist, emeritus professor at the University of Edinburgh,— Wikipedia.
129 *"Peter Higgs"* Wikipedia.   https://en.wikipedia.org/wiki/Peter_Higgs
130 See p.p 55-60

4. *The Linguistic Deconstruction of Science*
Extending the "poison" into the rest of science.

The *"quark"* linguistic deconstruction of nuclear particle physics consists of the replacement of objectively referenced scientific categories by artificially constructed word categories which are established solely by a consensus belief. This linguistic deconstruction requires the replacement of scientific "proofs" using well-designed experimental methods with "proofs" by a consensus of "experts."

This is not to say that the practitioners of artificial consensus categorization abandon empirical data altogether. Rather such data must be *"interpreted"* to *"fit"* the artificial consensus categorization. This often requires that the scientific method which identifies and tests significant *"cause and effect"* variables be abandoned. The technique of replacing *"cause and effect"* analysis with a consensus-supporting *"interpretive fit"* of data was pioneered by the 1968 Feynman deception.

Richard Feynman deserted the known *"cause and effect"* which identified accelerated electrons as scattering off nuclear material at x-ray orbital distances and reinterpreted the data to *"fit"* the *"quark"* artificial categorization.[131] The electrons were now said to be scattering off the nucleus, not at x-ray orbital distances. The *"softer scatter"* from electron orbital distances would be less energetic than a predicted "hard scatter" off the nuclear protons. The reduced energy of scatter was said to indicate the electron was scattering off a *"quark-like"* component with reduced mass rather than the whole proton with its full mass. Thus *"cause and effect"* analysis of the data was replaced by an *"interpretive fit"* of the data.

The Feynman 1968 distortion of data was implemented to further careers in academic science. The data re-interpretation was accomplished to support the artificial *"quark"* category which allowed the packaging of speculative particles proposed by academic science with the known and proved particles of nuclear physics. Fairly rapidly the technique of

---

131 SEE Chapter 2, p.p. 19-24.

removing scientific categories from their objective references and repackaging them as artificial categories which were established by consensus was extended to political causes. Specifically, the politically inspired *"environmentalist movement"* recognized a powerful new weapon in the implementation of distorted science.

By the 1980s, within a few decades of the rise of the *"quark"* deconstruction of nuclear science and its supporting Feynman data distortion, an assault was begun by an environmentalist consensus view against the published research discoveries of a well-known soil scientist, Edward C. Krug. Dr. Krug had graduated first in his class from Rutgers University under the mentorship of a world-renowned soil scientist, Dr. John Tedrow. In 1981, Krug was appointed a member of the National Acid Precipitation Assessment Project (*"acid rain"* project).[132]

Krug's *"acid rain"* project was a 10 year study appointed by the Federal Government to determine if *"acid rains,"* supposedly produced by coal-fired power plants, were degrading Northeastern forests and waters.

> *"[I]n the late 1970s. Scientists in the United States, Canada, and Scandinavia became alarmed at what they believed was massive environmental degradation caused by sulfur-dioxide laced rain that came from coal-fired power plants. The media followed with hundreds of apocalyptic stories, such as 'Scourge from the Skies' (*Reader's Digest) *'Now, Even the Rain is Dangerous'* (International Wildlife), *'Acid from the Skies'* (Time), *and 'Rain of Terror'* (Field and Stream)."*[133]

Krug's *"acid rain"* project studied the problem by taking core samples, primarily from Northeastern lake bottoms, comparing them with acidic lakes from other parts of the US and the world and studying the actual causes of forestry degradation in the target area. In 1983, Krug published the initial results of these studies in *Science Magazine.*

---

132 *"Acid Test: Edward Krug flunks Political Science."* Anderson, William, University of Tennessee, Chattanooga. Jan. 1992. Reason Foundation archives; Document #2304401
https://www.heartland.org/sites/all/modules/custom/heartland_migration/files/pdfs/5874.pdf
133 Ibid. p. 2 .

"The article, co-authored by Charles R. Frink, was entitled 'Acid Rain on Acid Soil: A New Perspective.' It was the only invited article by a NAPAP scientist to be published in Science.....

".......Krug and Frink looked at historical land-use patterns in the United States, Canada, and Scandinavia and concluded that existing soil chemistry was as important or more important to the pH of a lake or stream as acid rain. To perhaps oversimplify, it is acid in the surrounding plants and soil, not that in the rain, that causes acid lakes.

"Krug and Frink noted that acidity in lakes and streams had positive correlations with land use, a point verified by the EPA in 1989. Core samples taken from the bottom of many Adirondack lakes show increased acidity in the recent past but also show they were acidic and fishless before European settlement."[134]

Krug had found that, at the turn of the 20th century, Adirondack lakes had become less acidic, due to the fire clearing of surrounding forests, but were now returning to their pre-civilization acidic conditions.

"Trout survived better in the Adirondacks at the turn of the century than in earlier times because of slash-and-burn logging of that area. Eliminating the acid vegetation caused the soil to become more alkaline (a high pH), reducing the acid flowing into lakes and streams. In turn, the lakes became more hospitable to fish. After 'Forever Wild' legislation stopped the logging in 1915 the watershed reverted to acid soils and vegetation, and the lakes became acidic again."[135]

Krug *et.al.* had converted the term "acidic lake" to an objectively referenced scientific category by defining it as a body of water with a pH of 5.0 or less. To this category he had applied the hypothesis proposed by the *"acid rain"* project. Rains with high sulfur-dioxide content, such as those supplied by Northeastern coal-fired generators, were tested as the independent variable— as the cause of acidified Northeastern lakes. The hypothesis had to be rejected because the lakes were found to have been

---

134 *"Acid Test: Edward Krug flunks Political Science."* Anderson, William. Op. Cit.
135 Ibid. p.p.3-4

acidic before the arrival of civilization and the coal-fired plants.

A second independent variable was discovered by the Krug research as a significant causal factor for lake acidification. High acidity of surrounding vegetation and soil were found to be highly correlated with lake acidification. Lakes with high surrounding acidic conditions— but no "acidic rain"— were found to be acidic. In contrast, lakes with low surrounding acidic conditions— but which are subject to acidic rain— were discovered to be low in acidity.

In 1989 Krug was commissioned by the U.S. Department of Energy to publish an exhaustive assessment of acid rain theories and the *"acid rain"* project (NAPAP) research results; those *"objectively referenced"* studies of lake acidification which found adjacent soil and vegetation acidification to be the primary cause.

> *"The assessment went through peer review and was widely praised upon its publication in 1989. Erik Eriksson of Sweden, considered to be the father of acid-rain theories, sent DOE an unsolicited letter commending Krug's report as 'most welcome for shifting evidence from the sea of loose speculations often found.'"[136]*

What Sweden's Professor Eriksson failed to recognize was that his *"sea of loose speculations"* had become an alternative *"consensus category"* produced by a Wittgenstinian transformation.

For political reasons, the government's *Environmental Protection Agency* (EPA) had opposed Krug's 1989 assessment which he had based upon his objectively referenced scientific categories and subsequent analysis. They wanted the Krug assessment replaced by a "consensus assessment" made outside Krug's objective scientific referents. To this end, the EPA secretly convened a *"peer review panel"* to examine the Krug *"acid rain"* assessment.

> *"[T]he EPA, which was responsible for many of [Eriksson's] 'loose speculations,' secretly organized a 'review' of its own. Unlike the accepted scientific practice in which reviewers of*

---

136 *"Acid Test: Edward Krug flunks Political Science,"*p.p. 4-5 Anderson, William. Op. Cit.

differing opinions study the prospective manuscript, the EPA chose only those scientists who disagreed with Krug's organic acids theory.

"Keith Eshleman, a proponent of the mineral titration theory [the theory disproved by Krug's data], *lambasted Krug's work in a summary of the reviews, calling it 'highly misleading and oversimplified' and 'theoretically implausible and inconsistent with empirical observations.' Since the article had already been peer-reviewed and published, the secret reviews seemed to have no purpose."*[137]

But the secret peer review did have a purpose. Eshleman's critque of Krug is completely a Wittgensteinian linguistic transformation which removed Krug's work from all objective referents. Eshleman's claim that Krug's studies were *"inconsistent with empirical observation"* is a characterization made completely without example. No concrete case is ever offered showing this "inconsistency." The charge is made completely outside such objective references and is validated only by the alleged *"expertize"* of the utterer. It is completely a Wittgensteinian linguistic deconstruction of Krug's extensive research; one which removes Krug's methodology-based research from having any objectively-referred data to a new category which is defined by the pejorative opinions of alleged experts.

We know what the Krug data were which the *"secret peer review"* was meant to deconstruct. Krug had shown with his bottom core samples that Northeastern lakes had been acidic before the arrival of Europeans and before industrial acid rains. He had shown that adjacent acidic soils produced acidic lakes in the absence of acid rains and that lakes without adjacent acidic soils, but in the presence of acid rains, were not acidic. It was this data which needed to be removed from public consciousness and replaced with a deconstructed view of Krug's studies as *"inconsistent with empirical observation."*

The EPA used its Wittgensteinian deconstructed *"peer review"* to drive Krug from professional science. In late December, 1990 Krug took his case to the general public during an interview on the CBS News program *"60 Minutes."* In early January, the EPA responded to the Krug

---

137 Ibid. p.p. 6

appearance on CBS with a letter to the news organization. In the letter, Krug was said to *"have limited scientific credibility."* The letter goes on to quote the EPA's *"secret peer review"* of Krug.

> *"On the next page* [of the letter to CBS], *the EPA document gave selected damning quotes from the secret review, presenting them as the views of unnamed 'eminent' scientists."*[138]

The EPA sent copies of the CBS letter outlining the *"secret peer review"* of Krug and his work to several friendly press outlets and to environmentalist publications.

> *"The* Washington Post *printed a story on January 14, 1991, that told of Krug's 'limited credibility.' The same allegations appeared in other publications, such as* The Environmental Writer *and* Environmental Forum.*"*[139]

The linguistic deconstruction of Krug's studies among the lay public had a devastating personal effect upon his scientific career. He was driven from professional science and could no longer find employment. In a private conversation with Dr. Krug after the CBS debacle, he told the current writer that he had sent out over a thousand resumes which were all rejected by the scientific and academic agencies receiving them.

But Krug was not driven from science itself. A personal commitment to the objectively-referenced scientific method had lead him to understand the actual cause of lake acidification. By publicly stating this cause he had been attacked and driven from science. He began to recognize that most, if not all, environmental science had taken on the character of an artificial reconstruction of science which was driving out authentic science.

The current author had first encountered Dr. Edward Krug in the mid 1990s while the author was an editor for a small academic publishing house[140]. Paradigm publishing was monitoring those who had been driven

---

138 *"Acid Test: Edward Krug flunks Political Science."* p.p. 7. Anderson, William, University of Tennessee, Chattanooga. Jan. 1992. Reason Foundation archives; Document #2304401 Op. Cit.
139 Ibid.
140 The Paradigm Company, Boise, Idaho.

from science by the leftist politics which had begun to dominate government and academic institutions. At the time, Krug was the writer/publisher of a small monthly, titled *"Environment Betrayed,"* which Paradigm had solicited from him. Krug's newsletter subsequently became a book.[141]

Krug's *"Environment Betrayed,"* was documenting the corruption of the scientific method by which environmental science in general was suffering, and which had driven him, personally, from professional science. However, the newsletter was not being used as a weapon to gain restitution of his personal career. Rather, it was documenting assaults on all objectively-based environmental sciences, many outside his personal area of expertise, that of "soil science."

At the time, Krug did not recognize that the assaults upon the scientific method which he was documenting represented Wittgensteinian deconstructions of science. However, the current writer identified in Krug's documentations the application of the Wittgensteinian deconstruction of scientific categories which he had witnessed as a graduate student at Columbia University.[142] Specifically, Krug's description of the debasement of atmospheric science to support an alleged *"atmospheric ozone depletion,"* revealed a *"hard-data"* example of a Wittgensteinian deconstruction of objectively-referenced scientific categories to be replaced by a socially imposed pseudo-science.

Krug's example of the deconstruction of atmospheric chemistry became a core element in a book documenting the Wittgensteinian deconstruction of our culture. A portion of that book, which incorporates the crucial Krug data, is reproduced below: [143]

> *"That there is a Politics of Unreality abroad in the land is easy enough to prove. It is apparent, for example, in the environmental movement. Draconian regulations ultimately banning the use of freon in refrigeration have been forged at a huge social cost, especially to*

141 *Environment Betrayed-The Abuse of a Just Cause.* Krug, Edward C.   iUniverse, Bloomington, Indiana. May, 2012. ISBN: 978-1-4759-1126-8 (ic)

142 *"Why I don't have a Ph.D and, by extension— 'what does your Ph.D actually mean?'"* Dawson, L. The Snake River N Radiation Lab. http://www.paradigmphysics.com/phdat

143 *The Death of Reality (2015 Edition),* p.p. 7-22. Dawson, Lawrence. The Paradigm Company, Boise Idaho. ISBN: 978-0941995368 (Paradigm Company)

the poor. It is alleged that escaping freon from refrigeration units is destroying a stratospheric ozone layer. It is said that this ozone layer protects us from high frequency ultraviolet radiation from the sun, and that if it is lost we will be subjected to massive increases in skin cancer and other radiologically-induced disorders. This is presented as a 'collective threat,' and it is claimed that the 'selfish wills' of individuals to keep their personal comfort must be surrendered to the 'good of the whole.' This socialist idea is ALWAYS present in environmentalist claims, specifically that individual freedom of action must surrender to the 'good of the whole.'

"The ozone scare is especially poignant because, as we shall see, it clearly illustrates that appreciation of scientifically determined objective reality can be replaced by what might be termed 'politically generated formula thought.'

"The belief that stratospheric ozone is threatened by organic chlorine compounds (CFCs), such as freon, has been accepted as 'truth' by both the media and the politicians. Further, it is alleged that unless something is done about human destruction of the ozone layer a new plague will strike the earth. On November 4 of 1991, Newsweek told its readers, 'In April the U.S. Environmental Protection Agency announced...more cases of UV-induced cancer – an extra 12 million cancer cases among Americans over the next 50 years.' *In February of that same year, USA Today told the nation that a report from the United Nations Environment Program predicted serious health effects from ozone depletion. The newspaper approvingly quoted a Greenpeace spokeswoman as saying that* 'Ozone depletion is now so serious...that it now amounts to a threat to the future of all life on earth.'

"What are the scientific facts supporting these dire predictions? We find out there are no 'facts' as such. We find that the very idea of ozone depletion was politically generated, that the alleged 'cause' of this depletion has been changed to fit political needs of the moment, and that scientists have had research projects shutdown to prevent their results from contradicting the belief in 'ozone depletion.' In short, 'atmospheric ozone depletion' was a politically-generated formula not a factual discovery.

"The nature and history of the 'ozone scare' has been outlined by Dr. Edward Krug in his newsletter Environment Betrayed. *Krug is probably the nation's premier environmental scientist currently debunking green pseudoscience. In that capacity, he has appeared on the op-ed pages of the Wall Street Journal and on the CBS television program 60 Minutes. He has identified the origins of the 'ozone scare,' and it wasn't some relatively obscure scientist making a lonely discovery.*

"The 'ozone scare' began as a tactic invented by the environmentalist movement to stop production of the supersonic transport plane (SST) being contemplated in the early '70's. The SST was made the alleged 'threat' to the stratospheric ozone layer. First, it was said that water vapor in the plane's contrail would decompose to hydroxyl and deplete the ozone layer. When this didn't create the desired public reaction, a second SST 'threat' to the ozone layer was invented. Water was simply too 'natural' to make a good heavy in this eco-fiction. Krug states, 'The preconceived conclusion that SSTs are 'bad' was retained and the SST water vapor bogey man was replaced. The excuse this time— oxides of nitrogen (NO_2) emitted by high flying SSTs will erode the ozone shield (Johnston, 1971).'[144] *Ultimately, the nitrogen oxide 'ozone threat' succeeded in stopping production of the SST.*

"The greatly beloved American heretic Ralph Waldo Emerson once said that foolish consistency was the hobgoblin of little minds. The eco-movement would never be accused of being 'little minded' by Emerson's standards. It might even be said that they have managed to take 'big mindedness' to new heights of absurdity. They couldn't let the eco-fictions they had made of hydroxyls and nitrogen oxides die a decent death and slip ever so quietly into public forgetfulness. They actually resurrected the culprits as 'eco-heroes.' Green VP Al Gore's book, Earth in the Balance, *now called those nasty hydroxyls* 'natural detergents' *which cleansed the atmosphere. In 1984, with the newest CFC 'bogey' targeted on the ozone-depletion radar screen, Nature magazine characterized those once-killer oxides of nitrogen as 'defenders' of the ozone shield from the new CFC nasties.*

---

144 *Environment Betrayed,* Krug, E.C. June, 1994, Box 1161, Winona, MN 55987

"The greens weren't going to lose a bop, neat and keen racket such as 'ozone depletion' simply because they had knocked a multibillion dollar aircraft out of the sky. In the 1984 Nature *article, the US Academy of Science was cited to prove that increases in oxides of nitrogen gases in the atmosphere protected ozone from the CFCs.* '[Fortunately] atmospheric concentrations of...N_2O have been observed to be increasing. Continuation of these trends would delay the time when the dramatic effect of CFCs would occur.' *The article said that unless the oxides of nitrogen gases would increase faster than the chlorine from the CFCs, ozone depletion would increase dramatically. The same oxides of nitrogen which in 1971 were said to be 'threatening ozone' in SST exhaust were now described as 'defending' that same ozone against the new enemy CFCs.*[145]

"What's going on here? How can the same chemical compound be said to 'threaten' the ozone layer when the environmentalists wanted to ban the SST and then 'protect' that same ozone layer when the environmentalists wanted to ban freon? The answer, as Krug so well points out, is that the 'ozone depletion scare' has little to do with science and everything to do with the politics of fear. The ozone scare with its images of people's skin being fried in UVB light and becoming leprous with cancer worked so well with the SST that the environmentalists – who make careers out of 'protecting' us from dreaded monsters lurking in the unknown – simply could not give it up. They invented a politically self-serving thought formula that goes something like this: an incredibly thin and vulnerable ozone layer in the stratosphere is all that stands between us and dreaded radiation which threatens to turn the earth into burnt toast ('Ozone depletion is now so serious ...that it now amounts to a threat to the future of all life on earth' Greenpeace, 1991). *This vulnerable membrane about which you know nothing is about to be destroyed by (fill in the blank with a human activity you want suppressed and which can plausibly be demonized as ozone threatening). In short, the alleged 'threat to the ozone layer' was not discovered by objective research, it was invented and imposed by political forces seeking to profit by fear-mongering the idea. How else can you explain that the alleged "threat" to the*

---

145 Ibid.

ozone layer changed with such political expediency. The idea is a political unreality substituting for known scientific facts.

"We are not exaggerating when we call CFC ozone depletion a scam and a racket. Krug has documented that the popular belief in CFC ozone depletion is a myth which has succeeded only by suppressing scientific information. Those of us who are about to lose our freon-based refrigerators and air conditioners should, at least, be informed about the most significant of these suppressed facts. No less an authority than the head of the French equivalent of the EPA, Haroun Tazieff, tried to tell us about it in 1991. Tazieff said,'The Rowland and Molina theory [of ozone depletion due to CFCs] is unscientific because it is based upon a model of chemical reaction sequence without having proved the existence of the intermediary products; these reactions, which no one has ever reproduced in the laboratory, have never been observed anywhere.' *Did you get that? The way that CFCs are supposed to destroy the ozone has never been observed anywhere and indeed cannot even be reproduced in the laboratory. Freon is being banned at an estimated cost of well over a trillion dollars based upon a paper 'theory' of a chemical reaction between ozone and freon which has never been observed and which environmentalists have been unable to produce in the laboratory. How much more politically unreal can you get?*

"The paper theory was first put forward by F. Sherwood Rowland in an article in Science Magazine. *Dr. Arthur Robinson, himself an organic chemist currently conducting environmental research with the private* Oregon Institute of Science and Medicine, *explained how the Rowland thesis may have short circuited normal scientific controls. In a private conversation, Dr. Robinson said that Rowland's original article established what scientists term a "hypothesis." This consisted of equations for a speculative complex chain of chemical reactions which were, in Robinson's opinion, most probably mathematically sound. That is, the Rowland equations were consistent with known energy transfers and requirements associated with chemical reactions.*

"Dr. Robinson cautions, however, that there are many hypothetical chemical reactions which are coherent with the mathematical laws of

thermodynamics but which are reactions which don't actually occur. The fact that Rowland could provide mathematical feasibility for his reactions did not prove that they actually happened.

"The problem, according to Robinson, was that a hypothesis was elevated to the status of a 'fact' without going through the rigors of scientific testing. A chemical hypothesis is a speculation founded upon mathematical feasibility and guided by what Dr. Robinson called the 'intuition' of a well-trained expert in the subject. It must be submitted to experimental validation before it is accepted. Rowland's hypothesis was elevated to 'fact status' without experimental testing, most likely, in Robinson's opinion, by the lay press which saw the hypothesis' advantage to environmentalism.

"In science, an 'hypothesis' is defined as a factual proposition presented such that it can be disproved. Nothing reaches the level of hypothesis unless it proposes a method by which it can be disproved. Because Science Magazine *published Rowland's hypothesis without testing proves that a portion of established science had already deserted the scientific method in favor of a social consensus reality.*

"Dr. Edward Krug, whose newsletter Environment Betrayed *is an attempt to snatch authentic science from this voodoo of environmentalist unreality, has a knack for unraveling the science babble which the Greens use to mystify the common man. He points out that the amount of man-made freon which could possibly be 'blamed' for 'killing' the ozone is so minuscule in comparison to the volume of the atmosphere that a 'super killer' mechanism had to be invented. For every part of chlorine, the alleged 'killer,' from the CFCs there are 100 billion parts of air and 100,000 parts of ozone. To make the scare work, environmentalists had to make every chlorine atom 'kill' 100,000 molecules of ozone. They achieved this sleight-of-hand through Rowland's untested 'model' by which every chlorine atom is alleged to combine with ozone, breakdown the ozone molecule, then separate and commence the process all over again.*

"This is supposed to reoccur 100,000 times until the process is finally exhausted. Thus, one atom of chlorine could 'kill' 100,000 molecules of ozone.

"There was a slight problem with Rowland's untested hypothesis. When put to tests in labs it was proved to be wrong. The destruction of ozone it postulates could not be produced in the lab, nor had it been observed occurring naturally in the atmosphere. The hypothesis could easily be proved to be true or false by simple test. Reproduce the conditions in which the reaction is said to occur and see if it actually does. Create a stratospheric-like atmosphere with the proper amount of ozone in a sealed container, inject freon at the prescribed rate and photo decompose the freon. When tried, It did not 'kill' the ozone.

"Why? In answering this question, Krug is much harsher on the Rowland hypothesis than Robinson has been. Krug argues that the hypothesis is not only wrong, but that it is junk chemistry as well. The Rowland theory alleges that the 'super-killer' chlorine molecule – what scientists call dimers of chlorine monoxide must break apart at the point it is strongest. As Krug notes, 'Of the many ways that this chain of atoms [dimers of chlorine monoxide] could break apart, ozone theorists say that the chain breaks at its strongest links – the chlorine-oxygen bond. They say that the weakest link – the oxygen-oxygen bond – does not break apart! Imagine that!' [146]

"Dimers occur when two molecules of a compound weakly join together. They become something of a 'duplex' molecule and do not form a new compound because they can relatively easily be separated again. They are like magnetized ball bearings joining together. The ozone-depletion theory asserts that when two ball bearings are thus joined, one of the ball bearings will break in half before the two can be forced apart. We are being asked to believe that, if you strike two magnetically connected ball bearings with a hammer, one of the ball bearings will crack before the magnetic attraction breaks.

"The scientific method by which any hypothesis must be disprovable was simply abandoned. All authentic tests of the hypothesis had proven negative. Injecting artificial stratospheric atmospheres with freon – or even chlorine monoxide— did not 'kill' the ozone. Advocacy experimentation was substituted for hypothesis testing. Advocates of

---

146   *Environment Betrayed,* Op. Cit. July, 1994, p. 1

the disproved Rowland model spent time and money to see if they could produce a single anecdotal experiment to substitute for repeatable hypothesis testing. They spent time and money to see if they could force the molecule to break at its strongest bond using high energy. However, the pseudo scientists seeking a consensus for Rowland's now-disproved hypothesis could not even produce direct anecdotal data.

"Krug points out, 'No one has ever observed the chlorine monoxide dimer breaking at its strongest link and not at its weakest link.' *Unable to produce the reaction which the paper theory demanded, they did the thing that pseudo science must do in such politically sensitive areas. They went for a substitute, another molecule which they pretended was 'like' the molecule in question. Krug continues,* 'Indeed, the idea that this chain of atoms would break at its strongest link had to be inferred from the behavior of a different chemical molecule – $CLONO_2$. And, of the numerous experiments run on this proxy molecule, only once was it demonstrably seen that this strongest link – the chlorine oxygen (CL-O) bond-nearly always broke. Problem – This one favorable study cooked the sample with high energy lasers.' *After great numbers of studies on a 'substitute' molecule, the defenders of a discredited theory were only able to get the molecule to break apart at the predicted bond one time, and that with the infusion of massive amounts of energy. They had their needed anecdote.*

"Let us review then why your refrigerators and air conditioners are currently at risk. Freon is being banned because a hypothesis was presented by a political advocate which predicted that each atom of chlorine in that freon would 'kill' 100,000 molecules of ozone by a method which has never been observed in nature nor could be produced in the laboratory. It is asserted that we accept the politicized hypothesis as established 'truth' because one occasion produced one anecdotal piece of evidence. A substitute molecule could be shown to break apart in the predicted way after being bombarded with massive amounts of energy.

"How can this one experiment, obviously 'cooked,' as Krug calls it, stand in the stead of well established scientific testing principles?

Why would the breaking apart of a 'proxy' molecule under high energy bombardment be given more 'weight' than lab-controlled hypothesis testings which demonstrated that the chemical reactions predicted never occur under experimentally controlled conditions?

"Chlorine monoxide simply never acts as the theory's proponents claim, either in the presence or outside the presence of ozone. In a controlled atmosphere normal levels of stratospheric ozone are not reduced to oxygen by minuscule amounts of chlorine monoxide. The advocates of a politicized consensus science know this. It is a reason that the 'proxy' experiment replaces factual observation. The proxy experiment is useful because it helps establish an artificial consensus 'reality.' It is following the rules governing political unreality.

"In establishing belief in political, socially constructed unreality, one deals in the plausible, not in facts. One must make an artificial reality seem plausible in the same way that an alibi must be made 'plausible' while factual reality is ignored. The 'proxy' experiment described above was designed to shore up the 'ozone scare' theory's plausibility, not test its factuality. It is crucial to recognize the intentions of the advocates in these matters. The experiment, or at least its significance, is designed to overcome the psychology of disbelief. It is used to address a state of mind, not factual or external conditions.

"Here again is the essence of political unreality. One observation of desired consequence is given more credibility than the sum total of all natural observations combined. The will to believe is given more authority than either the objective witness of the senses or the rationality of the mind.

"Environmentalists have indicted themselves as deliberate practitioners of unreality by the fact that evidence which strongly and directly contradicts the 'ozone depletion theory' has been suppressed. During the scare over SST aircraft 'depleting the ozone,' the National Cancer Institute *actually established monitoring network to see if the allegedly thinning ozone would allow an increase in high-energy ultraviolet (UVB) to hit earth's surface. Even though the environmentalists succeeded in killing the SST program, the UVB*

monitoring system kept gathering data from 1974 onward. By 1985, when greens began arguing that 'ozone-killing' freon was going to fry us all, the UVB monitoring system had collected 11 years of data.

"Unfortunately, the monitoring data didn't support the 'ozone depletion theory.' Not only was there no increase in UVB radiation reaching the surface, but there was an actual decrease during the period the ozone was supposedly 'thinning' from CFC contamination. 'Ozone depletion theorists,' if they were interested in factual reality, should have wanted data on UVB hitting the earth's surface after 1985 since they had just 'discovered' the alleged 'ozone hole' at the South Pole, supposedly depleted through CFC contamination. Instead, the National Cancer UVB monitoring network was summarily shut down. Its data didn't help the 'ozone depletion' unreality and had to be suppressed. When the National Cancer Institute finally published its UVB data in 1988, environmentalists attacked the data, saying UVB rates were 'masked' by alleged 'air pollution.' Unfortunately, many of the sites, including the master site at Muana Loa observatory in Hawaii, were not subjected to air pollution.[147]

"Least anyone is trusting enough to believe that the elimination of the UVB monitoring system was not connected to a desire to protect the 'ozone depletion theory' from real-world data, consider what happened when a high government scientist suggested reestablishing UVB monitoring at the earth's surface to 'test' the depletion theory.

"According to the newsletter Inside Energy, *Dr. William Happer, the chief scientist at the* U.S. Department of Energy, *approached the office of Vice President Albert Gore to get aid in restoring the monitoring of UVB radiation. Gore, who is known as an 'environmental activist' and author of environmental books, had been warning the nation of the 'health dangers' from alleged freon poisonings of the stratospheric ozone layer. Happer, who had apparently taken the Vice President's words seriously, had thought we should resume the UVB radiation monitoring canceled in 1985.*

"Did the environmental vice president grab Happer's hand and shake it as a 'concerned scientist?' Not exactly. The 'Green' vice president who had once warned the nation of an imminent danger from a 'hole

147 *Environment Betrayed,* Op. Cit. June, 1994, p. 6

in the ozone' which he inaccurately predicted would open over the U.S.– fired Happer on the spot for making the suggestion.

"The ozone depletion theory and its voodoo chemistry must be accepted on faith and anyone suggesting that the idea be 'hypothesis tested' against real-world data was a friend of the damnable skeptics. Obviously, environmentalism is not concerned with an actual threat from ozone depletion. They have not rejoiced at unthreatening surface radiation levels, but have cursed the researchers for contradicting their carefully crafted artificial reality. They have made sure those measurements won't be taken again. The 'reality' of ozone thinning is what they say it is, not what if has been found to be. They believe their words create reality and are not committed to an objectively determined truth.

"But why? How can the environmental movement so rigorously desert objective science for favored belief? The answer, I'm afraid, is that that environmental movement is emotionally feeding the ideological appetites of its adherents and has little or no commitment to scientific objectivity.

"It has perhaps become a cliché to repeat that environmentalism is not conservationism. Environmentalists do not conserve natural resources to human use, they defend nature against human use. It is rooted in a form of prejudice, being for nature and against human activity. The movement began in unreality because the very words 'nature' and 'humanity' were given an artificial meaning of a very peculiar sort. One has to go back to the ancient pagan world to find an intellectual precedent. A kind of sanctity or sacredness was imposed upon the concept of 'nature.'

"This is not a wild exaggeration. The Colorado Sacred Earth Institute invited the UN regional director for the Environment, Noel Brown, to speak at a 1994 conference. Brown told the audience that the only way to prevent population pressures, pollution and energy consumption from destroying the earth, was to embrace what he called a 'new spirituality' which recognized the earth as sacred. The theme was also taken up by Al Gore in his book Earth in the Balance. *In one place, Gore reinterprets the Biblical story of Cain slaying Abel, asserting*

that the sin was not the killing of a brother made in the image of God, but the 'polluting' of the sacred earth with blood. In 1992, the U.S. Forest Service sponsored a Washington conference called the 'Spirituality/Wildland Interface.' Workshops included 'Symbolism and Spiritual Values in Experiencing Nature,' 'Sacred Land, Sacred Sex,' and 'Gaian Buddhism.' The 'sacred earth' concept has recently been baptized with the name of the ancient earth goddess Gaia. The earth as-goddess idea has even crept into 'science' with the 'Gaia hypothesis' which holds that the earth is a 'living organism.' This is treated as a 'serious scientific theory' in environmental circles.

"With the very concepts of the 'earth' and 'nature' being mystified, human activities which might impact the 'earth' and 'nature' also take on a novel artificial meaning. In the eyes of environmentalists, human objects and activities don't allegedly threaten scientific facts, they profane the temple. Pollution is not a practical problem, it is a violation of the religious sentiments. Unclean unbelievers have invaded the sanctuary and are trashing the Holy Order.

"In carefully looking at environmentalism, one can see the methods and origins of the phenomenon we are calling political unreality. The movement invests the natural order with presumed meanings, overlaying it with a religious sentiment. They have lost the capacity for thought because the world must fit into patterns and categories of their own making.

"Reality becomes what they have presumed it to be, not what it is discovered to be. They have overlaid the world with meanings of their own invention and call it 'reality.'

"Reality becomes what we presume it to be, not what it is discovered to be. The 'proof' for the presumed 'truth' is social opinion, not fact. Consider the fate of two men of 'science,' one who sought to pursue fact and the other who lent his weight to the 'popular' religiously-motivated opinion on ozone. The first was the Dr. William Happer mentioned earlier as the object of a vice-presidential purge because he suggested that the 'danger' alleged by the ozone depletion theory actually be tested. The second is Dr. F. Sherwood Rowland,'father' of the currently fashionable theory that ozone is

being 'killed' by your refrigerator.

"As we have noted, Happer was fired by the office of Vice President Albert Gore after the scientist had attempted to get Gore's aid in refunding an ultraviolet monitoring program which had been canceled in 1985. On June 17, 1993, an article revealing the Happer firing and its implications appeared in the Wall Street Journal *(Bad Climate in Ozone Debate). The revelation was 'refuted' by Michael Elroy, Chairman of Harvard's* Department of Earth and Planetary Sciences. *Prior to his firing, Happer had testified before a congressional committee and revealed the actual data by the 1974-1985 monitoring of UVB radiation hitting the earth. This was considered 'treason' by environmental enforcers since the actual data discounted the fashionable ozone depletion theory. In a letter to the Wall Street Journal, Elroy – with the full weight of Harvard 'earth science' behind him – discounted the controversy by ridiculing Happer.*

"According to this spokesman for establishment orthodoxy, Happer was deserving of his treatment because of alleged impure motives in attacking the credibility of revealed ozone 'science.' To reveal data which could disprove accepted 'truth' was considered a treasonable act. To ask for even more such data only emphasized the treason. Thus a scientist asking for data, the very object of the scientific method, was condemned as a heretic in one of the nation's leading forums by someone wrapping himself in the mantel of scientific 'authority.' Later, we will find out that much political unreality is sustained by 'political appointees' to high positions in the sciences and the university, political appointees who confer legitimacy upon favored untruths with their alleged 'authority.'

"Now contrast this treatment of Happer at the hands of 'authoritative' science with that given F. Sherwood Rowland. Rowland is the author of the 'super-killer' chlorine-atom theory, the dimers of chlorine monoxide which supposedly break apart in a way contrary to the laws of chemistry and have therefore never been observed to do so. As Edward Krug points out, 'Rowland's hypothesis predicts neither location, chemistry, nor rate of ozone depletion and Rowland's hypothesized reactions have been observed in neither nature nor

laboratory.' [148]

"Rowland was a chemist with the University of California at Irvine in 1973 when he invented the paper theory which was subsequently used to justify the draconian freon ban. His own attitude, as shown on a 1993 Discovery Channel *television program, indicated that he, himself, recognized his insupportable voodoo theory as a kind of career coup which would give environmentalists the pseudoscientific rationalization they needed to reignite the ozone-scare scam. He demonstrated that he trivialized the alleged "threat" to the ozone, but was enthusiastic about its impact upon his career. He said that when his wife asked him about his work, he replied,* 'It is really going very well. But it looks like we may see the end of the world.' *People who actually believe the world is about to* 'end' *don't glow enthusiastically about the prospect. Someone who sees a bright career prospect in making others believe it might.*

"Needless to say, Rowland has been paid handsomely for providing a 'theory' which can be sold the public as a 'plausible' ozone depletion mechanism. Both he and his absurd science have been the recipients of 'conferred legitimacy' by politicized 'authorities.' In August of 1994 it was announced that Rowland would be enshrined in the Smithsonian Science Hall of Fame. In 1995 Rowland was awarded the Nobel Prize in Chemistry.

"The Nobel Prize citation says of Rowland that he 'contributed to our salvation from a global environmental problem that could have catastrophic consequences.' This is the same 'environmental problem' which got William Happer fired for trying to measure. A member of the Academy which gave the award said it went to Rowland to put pressure on a forthcoming international meeting of the 'Montreal Protocol on Substances that Deplete the Ozone Layer.' Significantly, this same member of the Swedish Academy admitted that the award's 'prestige' was designed to overcome resistance from authentic science to Rowland's voodoo chemistry. Academy member Henning Rohde said, 'The Nobel prize will put a rest to this debate on whether the ozone hole really is a result of CFCs.'

---

148 *Environment Betrayed,* Op. Cit. Aug., 1994 p. 8

"You know, gentle reader, of what this 'debate' consists. Rowland's 'theory' is a failed hypothesis which failed in lab tests because it violates the laws of chemistry. Despite its authentic status as a disproved hypothesis, the Rowland theory was furthered in the pages of Science Magazine because it fit a socialist political agenda in environmentalism. To disguise Rowland's failure to meet scientific testing standards, an alternative one-time, anecdotal experiment was concocted using an 'alternative' molecule. To give credibility to the deception, scientific prizes were conferred, scientific prizes which were under the control of socialist political advocates.

"It is actually a debate between real science and an artificial science composed of politically imposed unreality. Science was surrendered to Rowland's unreality by the granting of a prestige award; The Nobel Prize.

"Truth will now be determined by a politically-constructed committee which will vote upon which particular 'truth' they prefer and confer an alleged honor upon it. As Krug states, politically-appointed scientific 'authorities' like the Smithsonian and the Swedish Academy are 'now passing off a disproved hypothesis as science fact.'

"The environmental movement has no need for truth and reality because they have THE FORMULA. The Formula is something akin to a fill-in-the-blanks complaint form, an easily recognizable model which tells them what to believe when someone fills in the blanks. The Formula goes something like this: Our precious environment, that most holy nature, is being de-sanctified by (fill in the blank) which is the profane man produced substance or the result of the profane human activity (fill in the blank) which will inflict the punishment upon us of (fill in the blank). The 'reality' is created when all the blanks are filled-in properly. An attitudinal predisposition by which all things natural are 'holy' and all things human 'profane' gives power to The Formula and makes it easy to determine that the blanks have been filled-in properly. It matters not one whit whether or if The Formula, as completed, makes a scientifically accurate statement. It is not evaluated scientifically. It is evaluated as fulfilling a religious prejudice, that of treating nature as sacred, and chastising man for violating that sanctity. The Formula, 'Our precious environment, that

most holy nature, is being de-sanctified by (ozone depletion), which is the profane man-produced substance or the result of the profane human activity of (the freon in our refrigerators), which will inflict the punishment upon us of (increasing skin cancer),' meets the criteria and is therefore 'true,' regardless of what the scientific evidence says.

"The point is that environmentalism uses an artificial mental formula of religious origins which is being imposed upon and substituted for factual scientific reality. It is a deliberate political unreality, in the sense that it is a system which is immune to factual correction [by the scientific method]. *"[149]*

---

149 *The Death of Reality (2015 Edition),* p.p. 7-22. Dawson, Lawrence. The Paradigm Company, Boise Idaho. ISBN: 978-0941995368 (Paradigm Company)

www.ingramcontent.com/pod-product-compliance
Lightning Source LLC
Chambersburg PA
CBHW060359190526
45169CB00002B/675